TIMBER INDUSTRY GHOSTS

TIMBER
INDUSTRY
GHOSTS

JEFF MOORE

AMERICA
THROUGH TIME®
ADDING COLOR TO AMERICAN HISTORY

Front cover:

Top: Former West Side Lumber Company 2-truck Heisler #2, construction number 1038, built in 1899 and displayed in the Tuolomne city park since 1960.

Bottom: Sawdust Burner at Willow Ranch, California.

Back cover:
Ghostly image of a McCloud Railway freight crossing the Lake Britton bridge in 2006 superimposed over the same scene taken a couple years later after the railroad had been abandoned.

America Through Time is an imprint of Fonthill Media LLC
www.through-time.com
office@through-time.com

Published by Arcadia Publishing by arrangement with Fonthill Media LLC
For all general information, please contact Arcadia Publishing:
Telephone: 843-853-2070
Fax: 843-853-0044
E-mail: sales@arcadiapublishing.com
For customer service and orders:
Toll-Free 1-888-313-2665

www.arcadiapublishing.com

First published 2019

ISBN 978-1-63499-138-4

Typeset in Minion Pro
Printed and bound in England

CONTENTS

ACKNOWLEDGMENTS

I've had the idea for a project like this for a few years now, but it took the gentle insistence of Fonthill's Jay Slater to get me to put aside enough time to make it happen. Thank you Jay for keeping after me until I said yes. Second, a big thank you to those who helped plug a few photo gaps, specifically Steve Moore, Tom Moungovan, Lee F. Hower, Jim Heringer, and Bruce Thompson. Third, thank you to those who helped with information or other aspects of this book, especially John A. Tuabeneck and Jack Botfield. Fourth, thank you to my dad and my late mother, who tolerated and promoted their son's interest in railroads and history from an early age. And finally, but certainly not least, a big thank you to my family for tolerating my eccentricities and generally not complaining too much when I pull off the road or deviate from established plans to get a photo or check some relic out and otherwise supporting these projects as they happen. I apologize if I left anyone out.

I'd also be very remiss if I don't thank you the reader for buying and reading this book. These efforts really aren't anything without you. I sincerely appreciate each and every one of you who reads these words.

All photos in the book are mine unless otherwise noted.

INTRODUCTION

Humans have always had complicated relationships with forests. The plants and animals making up forest ecosystems have been a major source of raw materials humans use to clothe, shelter, defend, and feed ourselves. At the same time, lands occupied by forests have been seen (and still are in many places in the world today) as space that could be used to graze domestic livestock or grow crops to feed ever expanding human populations. Dense forests also tend to trigger primal fears dating back to when our ancestors dreaded the large predators dwelling within them, and because of that forests are quite often closely associated in legends and lore with danger and dark forces. Certain elements of the human population have capitalized on this, using the relative protection of forests to hide illegal or illicit activities from prying eyes of disapproving others.

The most obvious way people have used forests has been in the extraction of raw materials for human uses. In less developed societies, fuel for cooking and heating fires is perhaps one of the widest uses of wood materials harvested from forests around the world. In the developed world, construction lumber and paper are the chief and most recognizable forest products. Sawmilling as an industry may have existed in France as early as the 1200s and was widespread throughout much of Europe by the 1500s. Shipbuilding to expand, defend, and transport the nascent empires constituted one of the principal uses of lumber in this period, and shortages of suitable timber in Europe in part drove the colonization of the rest of the world. More often than not, one of the first industries a colonizing power established in newly acquired territory lay in securing trees capable of producing either ship masts or shipbuilding timbers.

The development of the timber industry in what would become the United States commenced with primitive sawmills built in Jamestown Colony in 1625 and South Berwick Township, Maine, in 1631. Human muscles and water respectively powered these early sawmills, with steam power largely replacing both over the first half of the 1800s. The introduction of steam technology together with a growing railroad network allowed the timber industry to expand beyond the edge of the water, and the geography of lumber production rapidly wandered west and south from the New England states. Railroads in particular allowed vast quantities of lumber to be economically shipped long distances from new centers of lumber production to the swiftly building cities.

The forest products industry continued to rapidly expand and quickly became an economic, technological, and social force in and of itself. The high capital outlay requirements of sawmilling in increasingly remote locations initially favored giant corporations, and by the first decades of the 1900s the industry largely comprised of integrated companies that owned and controlled all steps of the process from logging through sawmilling to sales and distribution networks. A typical enterprise drew logs from large contiguous forest stands, through which company loggers living in isolated camps would ply their trade. The company typically built a dense network of logging railroads to bring logs to a centralized sawmill, usually but not always located in a town also owned and operated by the company. Finished lumber from the sawmill would be shipped to the outside world, usually by rail and occasionally over a common carrier shortline railroad owned by or closely affiliated with the company extending from the mill to a connection with a mainline railroad. These operations tended to remain in one location for several years to several decades, depending on available timber supplies. It was not uncommon for such companies to make cross country moves in search of new timber reserves as they ran out of existing timber, often bringing exotic place names and employees with them.

Technological advances—especially internal combustion engines and improving highways—from the late 1920s onward changed the face of the industry. The integrated companies relied on large volumes to cover their enormous operating costs, but cheaper cost structures associated with the new technologies allowed smaller companies to enter the business. The timber industry as a whole became fractured, as contract loggers and truck drivers—called gyppos—increasingly handled the woods ends of the business. The industry thus expanded into places in which bigger operations could not economically operate.

The U.S. timber industry has faced rapid and substantial changes in the last decades, forced by some combination of increasing environmental regulations, continued technologic development, competition from both other building materials and foreign lumber imports, and the reality that the industry cut trees far faster than forests could regenerate through the last half of the 1900s, especially in the Pacific Northwest. The U.S. remains one of the most predominant producers and consumers of lumber and other forest products, but the nature of the production end has substantially changed. One giant sawmill drawing raw logs from a vast geographic area in which dozens of smaller sawmills formerly operated have become the norm today.

The ever-changing face of the industry has left many ghosts in its wake. One form is the imprints and remnants of the vast infrastructure the industry once built, everything from logging railroad grades and trestles to now empty log ponds to foundations to parts of old sawmills. The second form this book will explore is logging equipment and technologies on static display. The third form is equipment and places the industry once used that have been repurposed. The last form is miscellaneous remnants that don't neatly fit in the above categories. This book by no means presents any form of comprehensive coverage of this subject, and it probably focuses too heavily on operations or geographical areas of particular interest to me. However, what's explored in this book is representative of timber ghosts that can be found almost everywhere this fascinating industry once operated.

1

ABANDONED IN PLACE

Among my paternal grandfather's photos from the late 1940s is a snapshot of cables wrapped around a large stump, most likely on or near the tree farm in the mountains west of Drain, Oregon, at which my grandmother spent the last years of her childhood. The cables are the remnants of a guy wire supporting a nearby spar tree once used to skid logs into a central landing in a previous logging operation, a common method employed to harvest the old growth forests on the steep slopes of the coastal ranges after 1920.

These cables are emblematic of the many relics the various aspects of the timber industry has left behind. Countless are the number of sites at which sawmills and other forest products plants of any size once existed; some of these are relatively intact, while others are marked by a few foundations, while many more have been redeveloped to shopping centers or housing developments that have completely obliterated all traces of the industry's existence. It's hard to find any forests that haven't been logged multiple times in the last century and a half, and while today's loggers take their tools and equipment with them, their predecessors did not always do so. It's not uncommon to find traces of early logging operations, ranging from lengths of cable to rail to sometimes almost intact pieces of equipment. Other ghosts of the timber industry past consist of things like old logging camps, abandoned railroad grades, trestle remnants, and other parts and pieces of long obsolete technologies and equipment.

Logging is and was one of the most dangerous occupations in the world. It's one thing to look at photos of early loggers at work. It's quite another to come face to face with the evidence of their labor in the woods, such as springboard notches ten or more feet off the ground on redwood stumps, the camps in which they lived, and the occasional intact steam donkey or other equipment they used and left abandoned. The first section of this book is devoted to some of these remnants as they appear on the landscape, largely at the spot at which they were last used, specifically to place them into context, explain their importance, and try to relate some of the stories they could tell.

NEAR DRAIN, OREGON. The photo that started this project, cables descending from a nearby spar tree wrapped around a stump sometime in the late 1940s. *David S. Moore photo, Jeff Moore collection.*

DRAIN, OREGON. One of the long-closed sawmills in the Drain, Oregon, area at which the author's great grandfather once worked, as seen through some of western Oregon's famous liquid sunshine. The Smith River Lumber Company moved the mill to this site from its former location west of Drain in 1961 and operated it until 1978. Only a few original mill buildings remain standing, while a lumber remanufacturing operation occupies part of the site today. *Steven D. Moore photo.*

SPRUCE MOUNTAIN, NEVADA. The mining industry faced a major problem as it spread into the Intermountain west. The smelting process required to extract metals from waste rock consumed vast quantities of fuel, which was only locally available in the form of wood. Unfortunately, wood tends to burn quickly and at temperatures insufficient for the smelting process, a problem compounded by the relative scarcity of trees near most mining districts. Pinyon Pine and Junipers tended to be the only available tree type species in the region, and they were concentrated at higher elevations.

The industry solved these problems by introducing charcoal production. This process as applied in the region mostly consisted of wood piled into a dome shaped structure which would then be covered in soil and leaves prior to firing; the burning process could last for up to a month. The resulting charcoal would then be transported, usually by wagon to area smelters. Several piles that were never fired still exist in remote parts of the Spruce Mountain Mining District in northeastern Nevada. *Bruce Thompson photo.*

DEATH VALLEY NATIONAL PARK, CALIFORNIA. In some areas of the Intermountain west, the mining industry imported mostly Swiss engineers to design and supervise construction of brick or stone ovens that reduced the charcoal firing process down to two weeks or less. Examples of these ovens can still be found scattered across Nevada and eastern California; perhaps the best preserved of these are ten ovens in the Panamint mountains in the western part of Death Valley National Park. The kilns date from around 1879 and were in use for only about three years producing charcoal for a silver smelter about thirty miles away. Coal and other fuels transported by railroads as they reached the region and the playing out of most mining districts largely ended this aspect of the timber industry by the end of the 1800s.

SAMOA, CALIFORNIA. Around 1893, high property prices on Eureka's waterfront prompted brothers Edgar and Silas Vance to purchase acreage on the peninsula separating Humboldt Bay from the Pacific Ocean due west of Eureka. The brothers had inherited substantial lumbering and railroad operations from their father, and the new site—named Samoa after the South Pacific Islands, then famous due to tribal wars—gave them the room to build the new modern band mill they desired. Construction also included docks, a new railroad extending north from Samoa towards the timberlands, a roundhouse and machine shop to house and maintain logging and railroad equipment, and a complete company town to serve all needs of the employees. The Samoa mill opened for business near the end of 1894.

NO SMOKING IN PLANT
(ONLY IN DESIGNATED AREAS)

LOUISIANA - PACIFIC

The Vance Brothers only operated the property for six years, as on September 1, 1900, they sold out to A. B. Hammond. Hammond already had substantial, mostly timber industry-oriented holdings in Oregon, Montana, Louisiana, and elsewhere. The Hammond Company expanded and modernized the operation under several different names over the next half century before selling out to Georgia Pacific Corporation in 1956. G-P added a stud mill, a plywood plant, and a pulp mill to the Samoa operations, but closed the logging railroad in 1961. In 1972, G-P included the Samoa operations in the portion of its assets spun off to the newly created Louisiana Pacific Corporation in order to settle complaints the Federal Trade Commission filed against G-P. Louisiana Pacific sold the paper mill and shut down all other facilities in 1995, but by 2008 the paper mill closed as well. Little remains of the sawmill save for concrete slabs and an occasional old sign.

SAMOA CONTINUED. The Port of Humboldt Bay now owns most of the sawmill site and other industrial properties. The old Hammond railroad roundhouse and machine shops still stand and today house the Timber Heritage Association collection of logging locomotives and other artifacts of the Humboldt County timber industry, and the group runs speeder trips on short pieces of restored railroad lines around the bay and one day hopes to start a tourist railroad operation. Simpson Timber Company acquired the rest of the Samoa townsite from Louisiana-Pacific along with the company's timberlands in 1998 and then sold it at auction in 2001. The town is presently owned by The Danco Group, a local investment company, which is in the process of redeveloping the town and starting the process of selling the houses, with the current occupants having the first opportunity to buy the house. Perhaps the best-known remnant in town is the famous Samoa Cookhouse, originally built to feed company employees but now offering a full logging camp meal service catering to the tourist trade. Other ghosts around town include the huge communal garage, of which almost every company town seemed to have at least one, and the "Samoa Block," a 24,000-square-foot building that once contained the company offices, store, and other facilities.

SLOAT, CALIFORNIA. The Sloat Lumber Company built the first sawmill at Sloat in 1912. The sawmill and associated facilities carried on under a succession of owners, several disastrous fires, and lengthy shutdowns during the Great Depression until the final owner, Sierra Pacific Industries, closed the mill for good in 1991. A flywheel that once guided one end of the band saw survived the destruction of the mill, and now on display holding a commemorative plaque next to what had been the mill's front gate. The wood frame in the background was once used to load log truck trailers onto log trucks as they left the mill headed back to the woods. A gravel company now uses part of the site.

DEE, OREGON. David Eccles and a number of close relatives and associates incorporated the Oregon Lumber Company (OLC) in 1889. In 1906, OLC built a new mill at Dee, Oregon, along with the Mt. Hood Railroad connecting the mill with the Union Pacific in Hood River. A logging railroad supplied logs to the mill until 1942, when trucks took over that role. Perhaps the biggest change made to the plant occurred in 1948 when OLC added a hardboard plant to the Dee facilities, originally making hardboard from bark and then sawdust and woodchips as the manufacturing process matured.

The Edward Hines Lumber Company of Chicago purchased OLC in 1955. Hines kept the sawmill going but had trouble keeping the hardboard plant current with the constantly changing market conditions for its products. Hines sold the operations to U.S. Plywood-Champion Papers on May 31, 1966. The new owners closed the sawmill but dramatically expanded the hardboard operations. Champion shuttered the plant in 1984, but a new group of New Zealand based investors reopened it as Dee Forest Products in 1986. The final curtain fell in 1996 when a spectacular fire destroyed much of the plant. Most of the site, seen here from a passing Mount Hood Railroad excursion train, now lies abandoned.

WILLOW RANCH, CALIFORNIA. Sawmilling at Willow Ranch, tucked near the very northeastern corner of California, started shortly after the Nevada-California-Oregon Railway pushed its line through the site around 1911. The timber industry in Willow Ranch remained small until 1926, when the Crane Creek Lumber Company purchased 194 million board feet of timber from the Modoc National Forest. Crane Creek initially built a box factory in Willow Ranch and a sawmill on Lassen Creek about five miles to the southeast. In 1928, the company built a railroad powered by a single two-truck Shay between the two plants and into the forests beyond that brought logs to the mill and then lumber down to the box factory. This arrangement lasted until July 22, 1929, when a Southern Pacific train ignited a fire south of Willow Ranch that swept north and east, destroying the sawmill.

Crane Creek subsequently built a new mill in Willow Ranch and extended the logging railroad further into the woods, eventually reaching somewhere around 16.5 miles in length. Trucks fully replaced the logging railroad about 1930, though the company left the line in place until 1934 and the Shay remained stored at Willow Ranch until finally scrapped in 1947. The Willow Ranch Lumber Company assumed the property at some point, only to close the mill in 1959. Nearby ranches continue to operate, but nearly all traces of the lumber industry are gone save for the sawdust burner (see front cover), concrete foundations, and a commemorative plaque mounted on the flywheel from the band saw.

WHEELER, OREGON. In 1895, Coleman H. Wheeler completed a sawmill on the shores of Nehalem Bay just south of Wheeler, Oregon. An extreme shortage of level land caused him to build the mill almost entirely over water on pilings driven into the bay floor. In 1913, John E. DuBois acquired substantial interests in the Wheeler Lumber Company and enlarged the property. The Great Depression claimed the mill in the early 1930s, though smaller sawmills sporadically used the site until 1970. Only these pilings remain today to mark the spot of the mill.

PONDOSA, OREGON. In 1926, the Grande Ronde Lumber Company closed down its sawmill in Perry, Oregon, located a few miles north and west of La Grande. Grand Ronde already had a site selected for a new operation thirty-five miles to the southeast on the banks of Big Creek, near the small community of Medical Springs. The company moved the sawmill to the new site, named Pondosa after the Ponderosa pine trees that dominated the area, throughout 1927. Construction included a company town surrounding the sawmill and the Big Creek & Telocaset Railroad, a new shortline railroad connecting the mill with the Union Pacific mainline at Telocaset. A logging railroad spread into the woods to bring logs to the mill.

Grande Ronde's parent company Stoddard Lumber Company merged the Grande Ronde company into itself in 1929, only to sell the Pondosa operations to the Collins Company two years later. The Collins organization got its start in Pennsylvania before spreading to the west. The Collins family carefully managed its company and its resources, traits for which they were rewarded with higher than normal employee loyalty. The Pondosa mill thrived under their tenure. As was normal, trucks replaced the logging railroad in the early 1940s.

After several prosperous decades, the Collins family decided to sell out. In February 1959, the Templeton family, acting through their Mt. Emily Lumber Company, acquired the Pondosa mill and town. The Templetons only really wanted the timberlands included in the deal, and the Pondosa mill cut its last boards on May 1. The new owners sold most assets at a grand auction held on May 5, and demolition of the site progressed quickly afterwards, helped a lot by a fire on June 20, 1959, that destroyed the mill, shop, dry kilns, and twenty houses.

Almost nothing of the sawmill remains today except for the log pond and a few foundations of the mill. More modern ranchettes dot the area around the mill. The grade of the Big Creek & Telocaset remains largely as it was in 1959, complete with a large number of ties the scrappers left behind when removing the rails.

HINES, OREGON. There isn't a merchantable tree in sight from the western fringe of Harney Basin, yet the presence of a warm spring bubbling out of the rocks two miles south of Burns, Oregon, intrigued Edward Barnes as he worked with the U.S. Forest Service on assembling the Bear Valley Timber Sale. In August 1922, the agency offered 890 million board feet of mostly Ponderosa pine timber on 67,400 acres of ground centered some fifty miles north of Burns. The sale came with several important conditions, including building eighty miles of common carrier railroad.

Lumberman Fred Herrick originally won the sale, and he purchased the spring and associated land through Barnes for use as a sawmill site. Unfortunately, a crash in lumber prices in 1924 dried up Herrick's construction capital and rendered him unable to fulfill the sale terms, and the Forest Service cancelled Herrick's contract in late 1927. The Edward Hines Company from Chicago won the sale when the agency offered it again in June 1928. Hines committed its vast resource base into completing what Herrick started, and the mill cut its first boards in 1930. The Hines name soon became associated with the town that swiftly grew around the new mill.

A network of logging railroads spread from Seneca, Oregon, out into the timber. The Oregon & Northwestern Railroad, a shortline owned by Hines, brought the logs fifty miles south from Seneca to Hines. The mill was somewhat unique in that the entire lumber manufacturing process occurred under one roof, from the time the log travelled up the slip into the sawmill to the point Union Pacific pulled boxcars of finished lumber out of the half-mile-long drying shed. Trucks started supplementing the logging railroad in the 1930s and replaced it altogether in 1967, though the Oregon & Northwestern continued hauling logs until 1978. The Hines company continued upgrading the plant as time progressed, including adding a plywood plant in the early 1960s.

Increasing economic and environmental pressures caused the Edward Hines Lumber Company to pull out of Oregon in 1983. The company sold the Hines mill to Snow Mountain Pine, who continued to operate the mill and added a laminates plant to the facility in the late 1980s. Timber shortages caused Snow Mountain to close the mill in 1995 and sell the laminates facility to Louisiana Pacific, who operated it until 2007. A small kitty litter plant and a recreational vehicle manufacturer tried to make a go in some of the old mill buildings, but failed or moved on.

Most of the sawmill facilities and buildings have now been scrapped, with the machinery sold to Russia for reuse there. A few buildings are used for vehicle repair and storage, mostly for local government agencies, and a few other small businesses operate in other remaining buildings, but the vast majority of the site lies abandoned. The tall concrete smokestack dominates the site.

GARIBALDI, OREGON. The Cummings-Moberly company from Texas built a sawmill in Garibaldi, Oregon, on the north shore of Tillamook Bay in 1918. The mill passed through the hands of the Whitney Company before lumberman A. B. Hammond—see the Samoa section in this chapter—purchased the mill. One of Hammond's first acts was to build a tall smokestack so that the mill emissions would no longer smother the town.

The Depression and Hammond's death forced the mill to close in 1935. A couple different companies used the site to operate a sawmill and plywood plant until 1974. Local entrepreneurs purchased the site and converted it over the next several years to The Old Mill RV Park and Event Center. Only a few outbuildings from the old forest product plants remain today.

Like Hines, the smokestack dominates the old mill site. It's seen here watching over an Oregon Coast Scenic Railroad crew assembling a work train; the railroad operates excursion trains out of Garibaldi using steam and diesel locomotives. The GP-9 is painted as a Holstein cow to honor Tillamook County's dairy industry, made especially famous by the Tillamook Creamery and its cheese products.

TUOLUMNE, CALIFORNIA. In 1899, the backers of the Sierra Railroad incorporated the West Side Flume & Lumber Company to log some 55,000 acres of pine they held in the Sierras. The company built a sawmill in Tuolumne, which became the eastern terminus of the Sierra Railroad after they reached town in 1900.

The original owners sold the properties to the West Side Lumber Company in 1903. The company built an extensive narrow-gauge logging railroad network through its holdings to bring the logs down to the mill, which made the company well known amongst the railfan community, especially in the 1950s. Unfortunately, the last log train rolled into town at the end of the 1960 season, and then a fire destroyed much of the mill while it was closed due to a strike in 1962.

A couple different parties, most notably Taco Bell founder Glenn Bell, attempted to turn the remnants of the sawmill and narrow-gauge railroad into a tourist attraction under the West Side & Cherry Valley Railroad banner, only to have those efforts fail for the last time in 1979. The 380-acre sawmill site passed through several owners before the Tuolumne Band of Me-Wuk Indians acquired the property in 2002. The tribe has spent a lot of money redeveloping the old mill site in the years since, and some of the only sawmill remnants are the sawdust burner and the old West Side Lumber office building. The building may not be long for this world, as in early 2018 the tribe was actively seeking permission to demolish the old structure.

SIMS, CALIFORNIA. The Central Pacific Railroad opened up a lot of north central California timberlands to development as it pushed its railhead north through the tight confines of the Sacramento River Canyon. One of the smaller operations existed at Sims, a few miles south of Dunsmuir. One source states Knight & Shelby's built the mill around 1902, while another names the involved company as the Northern California Lumber Company. The Sims Lumber Company ran the mill from 1908-1912, followed by a George Raish & Sons until 1918. Once again sources diverge, with one listing the Shasta Mill & Lumber Company as taking over in 1919, while another credits a Grizzly Creek Lumber Company under the direction of one B. J. Pye as processing 60 million board feet of timber harvested from Hazel Creek in 1920. At least a few of these companies operated a standard gauge logging railroad extending up to ten miles up Hazel Creek, though only a mile is reported to have existed in 1920.

The date operations ceased is not recorded but probably happened in the early 1920s. The Civilian Conservation Corp built a work camp adjacent to the sawmill site in the 1930s. A U.S. Forest Service campground occupies most of the site today, and the agency has developed an interpretive trail that loops past what's left of the sawmill and CCC camp. Remnants of the timber industry operations are primarily a rock foundation, a saw wheel, the faint imprint of the log pond, and the water tank off a locomotive that is resting upside down.

FALK, CALIFORNIA. In 1882, a group of landowners up the Elk River southeast of Eureka invited experienced lumberman Noah Falk to help them develop their holdings, and together they incorporated the Elk River Mill & Lumber Company. Noah's younger brother Elijah joined the effort, and together the Falk Brothers built the sawmill, which cut its first boards in 1884 and registered its first commercial shipment in 1886. The mill and lumber yards took up nearly all available space on a shelf above the Elk River, which forced the buildings making up the surrounding town named Falk up onto the adjacent hillsides. A logging railroad built up the Elk River brought logs to the mill, and the Bucksport & Elk River Railroad hauled the finished lumber eight miles down to the docks on the bay. Noah ran the company well until 1920, when he retired and sold his interests. The new ownership carried on until the Depression idled the facilities in 1930.

The company attempted a comeback in 1936, but the by now antiquated state of the facilities impaired the mill's ability to compete, and it closed for good within a year. The last residents moved out in 1944, and the lumber company scrapped out the logging railroad and most of the equipment just before the Bucksport & Elk River salvaged its line into Falk in 1951. A few squatters held on until around 1961, then after that only a caretaker on the fringe of the property remained until the early 1980s. The abandoned town started to gain some notoriety as the 1970s progressed, which caused the landowner to dynamite almost all the buildings in 1979.

The land under the Falk townsite eventually passed into the hands of the Pacific Lumber Company (PALCO). Falk lay adjacent to the Headwaters Forest, a 3,000-acre grove of old growth redwood that became the flash point between the timber industry and environmentalists when PALCO laid plans to log the trees in the late 1980s. Many years of face offs running from lawyers clashing in courts to direct confrontations between loggers and environmentalists in the woods ended with a 1996 agreement under which the company agreed to sell the tract, plus 4,000 acres of surrounding second and third growth redwood, to the Federal Government in exchange for cash and administrative reprieve on logging the rest of the company's timberlands. The agreement took a few years to work out, and in 1999 title to the Headwaters Forest—including the Falk townsite—passed to the U.S. Government's Bureau of Land Management (BLM).

Falk today is a quiet place, located a mile or so up a paved trail. The BLM has placed many interpretive signs and panels showing historic photographs and explaining the natural and human history of the site. Remnants of the town and industry that thrived there includes a few rotted timbers of the once extensive bulkheads built to protect the mill from the Elk River and two vintage dump trucks, the barn in which they were parked at the end long since having melted into the forest floor. Others include colorful lilac bushes marking the long-gone entrance to a house, a small shack, a few foundations, and various pieces of trash poking up through the thick vegetation.

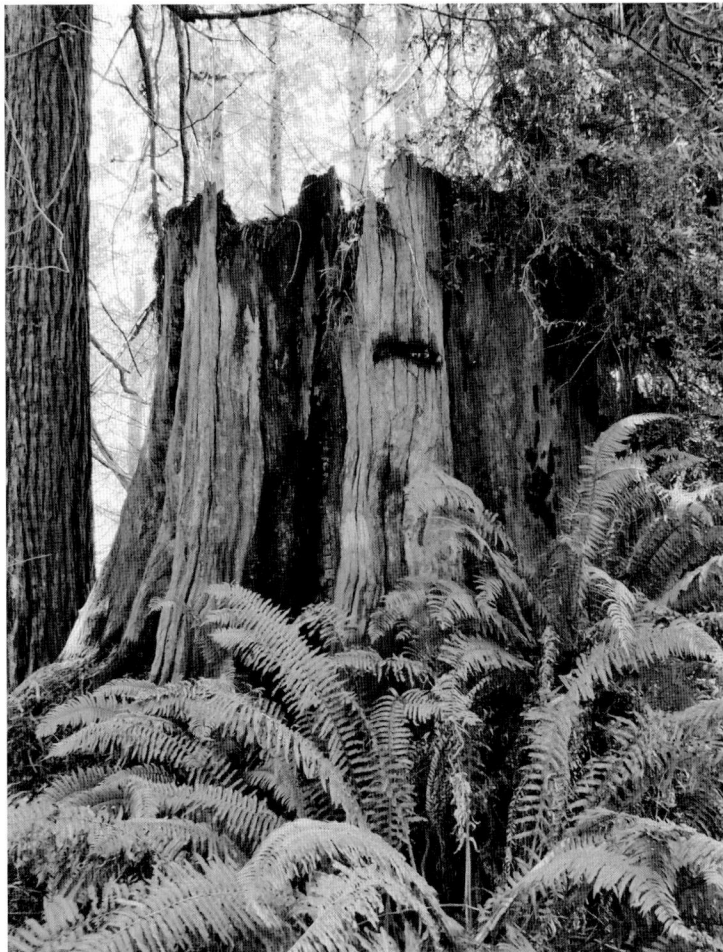

FALK CONTINUED. In most forest types, the stumps of logged trees tend to rot away after a few decades. A major exception is the redwoods found along the Pacific coast, where towering stumps remain standing a full 130 or more years after the tree had been cut. Redwoods have a generous butt swell near the ground, and loggers had to cut above that to get to straight grained wood; this forced them to work often 10-20 or more feet up off the ground, standing on springboards inserted into notches cut into the tree. Logging was and still is grueling work, and it would usually take a two-man falling team a week or more using axes and hand powered saws to cut down one of these giant trees.

The stump pictured here is one of many lining the trail into the Falk townsite described in the previous pages and is typical of many such ghosts of logging past scattered throughout the redwood region. Most of these stumps also have the next generation of redwoods growing out of them, either directly from the stump or from the surrounding root system.

REDCREST, CALIFORNIA. Lumbermen Bill Herndon and Don Martin built the Englewood Lumber Company mill in 1951. The mill passed through a series of owners and hosted a plywood plant owned by a third party from 1955 to 1967. At some point around the later 1970s, Eel River Sawmills bought the mill and operated it as their Englewood Division. The Eel River company dated from 1958, and the Englewood operation was a sideshow to their principle mill just north of Rio Dell.

Eel River Sawmills lived almost exclusively on U.S. Forest Service timber for decades, taking full advantage of a program that reserved a healthy percentage of agency timber sales for small companies like theirs. The company tried purchasing its own timberlands after a variety of factors reduced Forest Service timber sales to a trickle, but couldn't make the economics work out. The axe fell on the Englewood mill in 2000 and the Rio Dell mill in 2003. A subsequent owner bought the main mill but was only able to keep it going until 2005, when it finally closed.

The mill at Redcrest now lies mostly abandoned, though some economic activity is trying to take hold in a few of the old structures. An old mailbox sticks through a hole cut in a fence that reinforced the NO VISITORS policy, both next to a sign bearing the name of another vanished company. Inside the fence are bilingual signs promoting a safety culture to employees that are never coming back.

PRAIRIE CITY, OREGON. The Ponderosa Pine forests of eastern Oregon drew the interests of several lumbermen, especially as all-weather roads penetrated the country. The Edward Hines and Oregon lumber companies both built railroad networks towards the region respectively from the south and northeast, but a lot of merchantable timber separated the two large operations. Oregon Lumber's Sumpter Valley Railway reached Prairie City in 1910, but cut their line back to Bates in 1933.

The towns strung along the John Day River valley, especially John Day itself, became centers of this timber industry in the region. Prairie City had a couple small lumber operations during the years the Sumpter Valley served the town, then well after the railroad left a later group of lumbermen built the Prairie Wood Products sawmill, possibly the Blue Mountain Mills company in 1941. Prominent Oregon lumberman Don R. Johnson bought the mill in 1976 and subsequently expanded it, adding stud and planing mills to the small facilities.

Despite the odds generally arrayed against smaller mills like Prairie Wood, it survived until log shortages and falling lumber prices finally forced the D. R. Johnson company to close it in the fall of 2008. Unlike a lot of closed mills, the company has essentially mothballed the facilities, leaving the possibility of a resurrection at least open, should conditions allow for it to reopen in the future.

PINE, OREGON. Sawdust is an inevitable byproduct created any time a sawblade touches wood. The highly combustible material will very quickly accumulate in any sawmill or related industrial facility. Most sawmills would burn this waste in the boilers powering the mills, but that typically only accounted for a small portion of the sawdust created. Small mills would either pile the sawdust or burn it in open pits, but larger operations had to find ways to efficiently dispose of sawdust on a massive scale.

Steel lined structures built for the sole purpose of burning sawdust first appeared towards the later part of the 1800s, and the U.S. Patent Office issued its first patents for enclosed sawdust burners in 1888. The number and types of burners on the market multiplied in subsequent years, and they became a prominent part of almost every plant sawing wood. The most common type of burner in the industry became known as the "Wigwam" burner.

Pine, Oregon, located immediately south of Halfway and thirty-seven miles east of Baker City, is typical of many former sawmill sites where the burner is the only remaining trace a sawmill ever existed at this location. The Stil-Van Lumber Company, previous operators of a number of mills in the mountains east of Prineville, built this mill in 1949. The Ellingson Lumber Company added the mill to its collection of mills scattered throughout eastern Oregon in 1963 and closed it in 1972.

DORRIS, CALIFORNIA. In the end economics played almost as big a role in the demise of the sawdust burner as increasingly stringent air quality regulations. Lumber prices did not come close to keeping pace with skyrocketing raw log costs, especially in the 1960s, which forced the industry as a whole to develop new products that could be made from sawdust and other waste. Pulp and paper mills converted woodchips to useable products, and several existing timber companies built such plants to utilize the waste from their mills. Other companies, most notably Fibreboard, built paper mills and then bought existing sawmills so as to guarantee a stable supply of woodchips. Other uses for wood waste included co-generation plants that burned sawdust to create electricity, particleboards and hardboards and other similar glued products, and the like.

The use of sawdust burners all but ended by the 1970s, yet many can still be found, some standing silently by a still-active facility, while others such as the one on the previous page are the last remnant of a long-gone sawmill. These two old burners stand in Dorris, California, once home to a half dozen or so sawmills but now down to a single small moulding plant. The larger of the two presides over the Butte Valley High School football field and has been partially painted in that school's colors, while the smaller stands next to a closed millwork or moulding facility not too far away.

NEAR SATTLEY, CALIFORNIA. The Turner family operated numerous small sawmills at various points around Sierra Valley. Around 1902, the family built a new mill they called the Sunset Mill about three miles north of Sattley, plus a planing mill across the valley in Loyalton. Harry A. Turner operated the Sunset Mill until 1912, when he became involved in establishing the Sloat Lumber Company and moved the operations there.

The author has been unable to find much about the subsequent history of the site. A massive forest fire in 1924 is reported to have destroyed the old Turner mill and nearby buildings, including a shingle mill and machinery. A 1955 map of Sierra County identifies the site as "Turner Mill." The sawdust burner and other remnants at the site today suggest more recent sawmill activity.

SUNSET, CALIFORNIA. In 1948, Gayle Morrison and two partners incorporated the Sunset Moulding Company. The new company built a moulding plant at Sunset Landing, just south of Live Oak. The Morrison family bought out the partners in the late 1970s and has successfully operated the company since, which has now expanded to include additional facilities in California and one in Oklahoma.

The forest products industry as a whole became subject to intense public scrutiny in the 1970s over impacts to air quality these burners caused. As noted, the manufacturing end of the industry had generally by that point started selling their waste products instead of burning it, and most plants demolished their sawdust burners so as to avoid any perceptions they were still in use. The Sunset Moulding plant is an exception, as its burner still stands along the edge of their property. *Lee F. Hower photograph*

PONDOSA, CALIFORNIA. The dangerous and often remote employment opportunities the logging industry offered typically attracted a mostly single and highly transitory workforce in the earliest decades. These demographics started changing in the middle 1920s, which prompted the more alert and progressive companies to rethink their operations. Typical of these is The McCloud River Lumber Company, which abandoned its use of mobile logging camps in favor of an established semi-permanent communities for its woods-based employees. The company first employed this concept in late 1919 or early 1920 at a point they named Pondosa, which like the Oregon town of the same name is a corruption of Ponderosa pine. The first Pondosa existed until 1927, when the company moved the entire camp, including the name, ten miles south.

The McCloud River company operated the second Pondosa camp from 1927 until 1964, when corporate successor U.S. Plywood shut the camps down and moved all loggers into McCloud. Pondosa lay in the center of an 87,000-acre tract owned by The Red River Lumber Company but logged by McCloud under contract. The community consisted of a large maintenance shop, a store, recreation hall, cookhouse, and other facilities. The loggers and their families lived in the camp year-round, and the camp included a school through the eighth grade, with high schoolers bussed to McCloud.

Management of the site reverted to the Walker Family, who formerly owned Red River, after McCloud pulled out. Some buildings have been torn down and others have collapsed, but large parts of the camp remain intact more than fifty years after it was abandoned. Some of the interesting remnants include an old stone fountain in what had been the center of the camp and several family houses, some built on site and others such as this one assembled from old cabins left over from the mobile camp days.

PONDOSA CONTINUED. Centerpiece of the McCloud River camp in Pondosa were two buildings, the company store and the recreation hall. The McCloud River Lumber Company had a strong paternalistic attitude towards its employees, but that still did not prevent the company from requiring them and their families to purchase everything they needed through the company store. The principle store was located in McCloud, with the stores in the camps like Pondosa operated as satellites of the main facility. The stores kept all the needed basics in stock, and anything they did not carry could be ordered through the stores. The daily log trains operating out from McCloud and in later years trucks would bring supplies to the camp stores.

The recreation or club hall sat next door to the store building. It had a single large room that served many functions. Pool tables would be set up on occasion. The Siskiyou County public library would bring a selection of books down from the main library in Yreka one day a week. A man named Olsen would show movies every Wednesday night, and another man named Stedman operated a snack bar and game stand for the local youth. The building later also served as the union hall after organized labor made its way into the McCloud woods. The buildings still stand empty today; seen here are the front of the recreation hall and rear of the store building.

HAMBONE, CALIFORNIA. As noted on the previous pages, the McCloud River Lumber Company's Pondosa camp existed at two places. The original site after the camp moved in 1927 became known as Old Pondosa and then Hambone, reportedly after a ham bone nailed to a tree marking a water location on a wagon road that passed through the area in the late 1800s. Hambone had quite a history after the logging camp left; it started in 1929 when the McCloud River, Great Northern, and Western Pacific railroads chose the spot as the official interchange point between the three roads. The Great Northern owned the thirty-four miles of track extending from Hambone east to their mainline, over which the McCloud River operated freight trains. The Western Pacific had trackage rights into Hambone, but never used them, preferring instead to pay the other two roads to move their traffic between Hambone and their interchange with the Great Northern near Bieber, California. The GN stationed several employees in Hambone for several decades to officially process cars through the interchange until automation and centralization eliminated their jobs, plus a section gang to inspect and maintain the tracks. Two later sawmills operated at the site, an Elkins Cedar Mill that cut mostly incense cedar for the pencil industry from 1934 until 1944 and a smaller stud mill in the late 1940s/early 1950s.

Hambone lay largely abandoned by the end of the 1950s. A line change completed in 1956 bypassed most of the site, and all of the remaining structures were later burned. The last train rolled through Hambone near the end of 2003, and by 2008 the tracks themselves had been removed. There are many remnants of the entire history that can still be found, including a few still standing poles of the phone line the McCloud River Railroad built and many ties from the long-gone yard tracks. Remnants of the original camp include the base of the oil tank surrounded by spilled oil and part of the concrete slab upon which the maintenance shop once stood. An old section shed on the final railroad alignment is the only still standing structure from the original camps.

SOAP CREEK PASS, CALIFORNIA. The Pickering Lumber Company pushed its logging railroad through Soap Creek Pass around 1930. The site saw relatively little activity until the middle 1950s, when Pickering abandoned its logging railroad beyond the site and established a large community for its loggers and railroaders. Due to its relatively late construction Soap Creek was perhaps the most modern logging camp ever built.

The last Pickering log train departed Soap Creek at the end of the 1963 season. The logging part of the camp continued on for several more years until it too closed. Pickering's successor Fibreboard Corporation conveyed the camp to the Boy Scouts of America, who operated it as a summer camp for a few years before replacing it with an entirely new camp constructed a few miles away.

Soap Creek Pass lies mostly abandoned today. The old union hall is the most intact building in camp and is still used by the Boy Scouts to warehouse a few camps supplies. Most remaining buildings are in some stage of collapse, such as the shower building for the single men's part of the camp as seen here. Remnants of the railroad include the base of the water tower and the boards Pickering master mechanic Albert Botfield had laid in the old locomotive servicing area to ease effecting field repairs on the rail equipment.

KINYON, CALIFORNIA. The transition from mobile to permanent logging camps posed a transportation problem as logging operations moved progressively farther from the camp. Logging companies developed various methods for shuttling loggers to and from the woods. The McCloud River Lumber Company shop in Pondosa built one of the more unusual cars for this service around 1940. Master mechanic Vic Larsen designed and supervised its construction, starting with an old Caterpillar "Sixty" tractor and building outwards from there. The car, named the *Red Goose* by those around it, was largely despised by those who had to operate or ride it, yet it did what it was designed and built to do.

Crew vans and improved roads relegated the *Red Goose* to storage by around 1950. However, in 1951, the McCloud River Lumber Company built its last logging camp, named Kinyon after its superintendent. Logging operations based out of the new camp occurred in some rough terrain, which gave the car a new lease on life for a few years until the road network could be improved. The *Red Goose* was probably last used in the middle 1950s. U.S. Plywood didn't know what to do with it when the closed and demolished Kinyon in 1964, and it has been slowly deteriorating in place ever since.

SYCAN, OREGON. Timber industry giant Weyerhaeuser started buying timberlands around Klamath Falls, Oregon, in 1904. The company did not start building its sawmill until 1928, shortly after Great Northern reached the town, and when the mill opened in 1929 it had an output greater than the combined production of the next four largest mills in the region. Weyerhaeuser focused its initial logging activities on its holdings in the mountains to the west. Preparations to log its timberlands east of the city started in 1940. The west side logging railroad operated closed in 1956, while the east side operation lasted until 1990.

The mountains around Klamath Falls can get substantial snowfall, and Weyerhaeuser bought or built three of these spreader snowplows in order to keep its railroads open through the winter months. When Weyerhaeuser scrapped the railroads out in 1992 and 1993, it left one of the plows sitting on a few lengths or rail in what had been its yard adjacent to the old railroad shops at Sycan, though as this book was going to press, the Collier Logging Museum near Chiloquin acquired this plow and moved it to their grounds. The other two plows also still exist, one on display at Olene a few miles east of Klamath Falls and the other on display on the grounds of Train Mountain near Chiloquin.

McCLOUD, CALIFORNIA. The McCloud River Railroad and McCloud River Lumber Company operations also lay in some areas with often prodigious snowfall amounts. The McCloud River Railroad, early in its history, bought a large bucker plow from the Central Pacific Railroad; the plow worked well, but then came two especially hard winters in a row that overwhelmed the plow. In the winter of 1913-1914, the McCloud River Railroad car shop built a second plow patterned after the first at a total cost of $2,028.06 ($1,041.82 labor, $150 trucks, $12.95 Simplex coupler, $124.94 steel and iron, $58.11 hardware, $5.38 paint, $11.19 drawhead castings, $7.36 plow shoes, $417.62 lumber, $42.42 body bolsters, $62.50 wheels and axles, $69.45 miscellaneous and coal, and $24.34 store expenses).

The plow, McCloud River #1767, served the railroad well. The car shop rebuilt both of the plows in 1932 to provide greater clearance. The plow's wooden construction and age caught up with it in the winter of 1968/69 when it received substantial internal damage while trying to open the line after a series of storms dumped twenty feet of snow in some places. The plow sat on display at various places around McCloud for several years before the railroad finally moved it to its equipment storage yard east of town. Title passed to a private party in 1986; however, the entire rear of the plow fell off in the winter of 2006/07, and the doghouse on top of plow deck has collapsed entirely into the body since the date of this picture. Ironically, the original plow is still intact and in much better condition, on display at the Railroad Park Resort south of Dunsmuir.

PONDOSA, CALIFORNIA. Snowplows and railcars aren't the only pieces of logging equipment left abandoned in the woods. Concerns over damage to forest soils and young trees, caused by logs being drug by tractors as they became common in the woods, caused the logging industry to develop arches that would lift the leading end of logs off the ground. The Hyster company of Portland, Oregon, became the leading manufacturer of logging arches, many of which can still be found scattered throughout timber country. Some are on display in museums or municipal spaces, while others are abandoned around old sawmills or equipment yards. This one was placed on display next to the entrance of a small sawmill in Pondosa, California, originally built by the Harry Horr Company in 1930 and later operated by the Ben Cheney Lumber Company and then Louisiana Pacific. The sawmill, closed for the last time in 1980, now lies mostly abandoned, with the old arch now almost entirely enclosed by a thicket of young trees.

TUOLUMNE COUNTY, CALIFORNIA. Among the rarest of lumber industry ghosts in the woods are abandoned pieces of large equipment, especially items like donkey engines. The various hoists and similar machines used to load and unload ships along Eureka's waterfront planted the seeds for the donkey engine into lumberman John Dolbeer's head, and he filed patents for his creation in 1881. Skidding logs out of the woods prior to this time had been a herculean task, especially in the redwood country, and usually accomplished by teams of oxen or horses. The donkey engine allowed for logs of almost any size to be quickly and efficiently extracted from the forests.

The Willamette Iron & Steel Works of Portland, Oregon, was one of the primary manufacturers of donkey engines. Internal combustion equipment largely replaced steam donkeys by the end of the 1930s, though a few remained in service into the very early 1950s. Most logging companies scrapped their donkeys, but a relative handful that were too costly to remove were abandoned where they last worked. Scrappers caught up with most of them, especially during the World War II scrap drives, while others have been rescued for display. A select few still sit in the woods today, such as this pair of Willamette 12 x 14 Humboldt Yarders perched precariously on a hillside in the Sierra Nevada Mountains. They are Willamette construction numbers 1768 and 1858, purchased new in 1920 and 1922. By all appearances Pickering Lumber was in the process of moving these two machines to a new setting at the end of the 1930 season, only to have the Depression intervene, and by the time the company reopened in 1937 it had transitioned entirely to tractor logging. The two machines still exist and are remarkably intact almost ninety years after their last use in the woods.

TUOLUMNE COUNTY CONTINUED. Far more common than intact donkey engines abandoned in the woods are the sleds upon which they rested. Almost every donkey engine not affixed to a railcar or some other equipment would be mounted on a sled composed of two large logs, usually with the bottom of the front end carved off as is normal for most sled runners. It was not uncommon for a donkey engine to go through several sleds during its operating life, and often a worn-out sled would be left in the woods after the donkey engine had been removed. One such donkey sled on the old Pickering Lumber operation is seen here; a few parts have been cut away by scavengers, likely to retrieve metal parts, and the size of the cedar and pine trees growing up between and adjacent to the runners gives some idea of how long the sled has been sitting here.

Another remnant of the Pickering Lumber operations is this cab of a 1948 Peterbuilt log truck, left abandoned in the old truck boneyard near where the truck shop once stood at Soap Creek Pass. Log trucks started supplementing logging railroads in the first decades of the 1900s and eventually replaced them all together. Pickering purchased a large number of these trucks for off-highway use, especially in bringing logs from the woods into reloads established along the railroad.

BURNEY, CALIFORNIA. Most forests logged during the railroad logging area are littered with old railroad grades. Railroads were and are extremely costly to build, but most logging railroads would be in place for a few weeks to a few months and would be removed as soon as the trees had been cut, unless the company needed the tracks to move logs from more remote stands.

Rails tended to be removed from logging spurs as soon as practical, as they represented an enormous investment by themselves and would be reused in the next spur built into the next stand of trees. Some of the few exceptions to this are the short lengths of rail embedded in the pavement of Mountain View Road, known locally as Old Dump Road, a little less than a mile east of Burney Jr. Sr. High School. McCloud River Lumber Company logging spur #700 crossed the road at this point from 1955 until corporate successor U.S. Plywood shut the logging railroads down in late 1963 or early 1964. The old grade is now a road, and only these rails left in the road speak to the railroad that once existed here.

CRESCENT CITY, CALIFORNIA. A number of railroads projected routes connecting Crescent City, tucked in the far northwestern corner of the state, with the outside world, but none of them ever came really close to fulfilling those goals. This did not stop a healthy redwood lumber industry from developing in the area, and it grew to support a modest local railroad network. The Hobbs, Wall & Company built a large mill in Crescent City about 1871, and within a few years the company broke ground on a logging railroad running north of town.

About 1912, Hobbs-Wall completed its logging north of the city. The company abandoned its northern line and started building a new railroad to the south, eventually incorporated as the Del Norte Southern so as to have the power to condemn a right-of-way through an otherwise uncooperative landowner and be prepared for any railroad that might reach the area from the outside world. The Del Norte Southern mainline eventually extended out about ten miles, and several branches tapped timber stands off of this line.

Sources list the end of the Del Norte Southern as coming sometime between 1935 and 1940, when Hobbs-Wall exhausted its timber reserves. The Hobbs-Wall mill site is now occupied by the Safeway shopping center in town. One of the few remnants of the company's railroad system are a few still standing bents of a trestle over a gulch not far off Highway 101, a few miles south of Crescent City.

TUOLUMNE COUNTY, CALIFORNIA. Typical of most temporary log spurs is this one, built by Pickering Lumber Company most likely sometime in the later part of the 1920s. The builders did minimal to no grading, and appear to have laid the tracks more or less directly on the forest floor. The company took only the rails when it abandoned the spur, leaving almost all other track materials scattered around on the ground. Other remnants at the site include the donkey sled on a previous page and substantial amounts of cable. A later logging operation, most likely in the 1970s, scattered the ties around, and at least one logger took the time to drive a spike into the top of a stump.

ALGOMA, OREGON. One of the most impressive features used by numerous logging railroads is the incline, in which a logging spur would be built straight up or down the side of a hill or mountain. Logging railroads used these to rapidly gain or lose elevation, and thus save many miles of railroad construction. Specialized donkey engines would be used to raise and lower log flats and other equipment up and down the grade.

The Algoma Lumber Company built this incline in 1915 near their mill on Upper Klamath Lake. The incline was 2,600 feet long, gained 800 feet in elevation, and had a grade of fifty-seven percent. The company used the incline at first for only two years, as by 1917 they exhausted their timber in the area. Algoma rebuilt the incline around early 1927, mainly to harvest substantial timber burned the prior year, plus some additional stands the company purchased. This operation continued until sometime in the late 1930s when Algoma sold the remaining timber above the incline to the Kesterson Lumber Company. Kesterson continued using the incline to bring logs down, with the logs loaded onto trucks at the base of the incline after a levee break flooded the valley in 1940.

Kesterson exhausted the timber above the incline by early 1942 and scrapped it out for the final time. By this point the Algoma company was nearly out of timber to keep its mill located about two miles away from the bottom of the incline running, and it closed in 1943. The scar from the incline is still obvious, and a few concrete foundations remain to mark the old mill site.

LEAF, CALIFORNIA. Abner Weed had invested twenty disappointing years in the northern California lumber industry before he first encountered the near constant wind gusts howling off the northwestern flank of Mount Shasta that he deemed would be ideal for drying lumber. Weed secured a mill site, incorporated the Weed Lumber Company, and broke ground on a sawmill and surrounding company town. The Weed company started building a logging railroad about 1903, heading generally northeast from the new town.

The Southern Pacific closely watched the advance of Weed's railroad, as it mirrored their plans for a new line running towards Klamath Falls. SP officially acquired Weed's mainline in 1906 and extended it to the northeast, reaching Klamath Falls in 1909. As part of the deal the SP agreed to build at its expense connections to any Weed Lumber Company log spurs for up to forty miles out of Weed. All told Weed and corporate successor Long-Bell built about twenty-six logging railroads connecting with the SP through the years.

The longest lived of these departed the SP main at a point named Bray Wye and later renamed Leaf, just shy of thirty-seven miles out of Weed. SP installed the connection in 1911, and the spur gradually grew into a mainline extending eighty miles to the east, plus hundreds of miles of branches and spurs. The connection lasted until 1956 when Long-Bell scrapped the railroad for lack of any more timber to haul. These decaying ties are all that remain of the yard tracks at Leaf.

EUREKA, CALIFORNIA. Countless sawmills once ringed the shores of Humboldt Bay, and nearly every drainage cutting back into the surrounding mountains had a logging railroad built up it at one time or another. Flanigan, Brosnan, and Company built the Bayside Mill towards the south end of Eureka's waterfront in 1876, initially processing timber logged on Jacoby Creek north of the city and rafted down the bay. By 1909, new owners shifted the logging to new holdings located up the Van Duzen River south of Eureka, with the Northwestern Pacific Railroad handling the logs down into Eureka. In 1929, the Humboldt Redwood Company purchased the mill, and the Hammond Lumber Company (see Samoa) immediately bought a half interest in the company. The Depression caused the mill to close in 1932, at which time Hammond acquired full ownership.

Hammond reopened the facilities as their Bayside Mill in 1937. Though dwarfed by the main Hammond facilities in Samoa, the mill remained a small but important part of the company's operations up to the point Georgia Pacific took over in 1956. GP didn't like the idea of having two essentially duplicate facilities in the same neighborhood, and they closed the Bayside mill in 1960. Today the most substantial relic of the mill is the log dump built out over Humboldt Bay, which amazingly enough still has rails on it. Looming across the bay in the background is the long-closed GP paper mill in Samoa.

EUREKA, CALIFORNIA. John Dolbeer, noted above for receiving the first donkey engine patents, and William Carson organized the Dolbeer & Carson Lumber Company in 1864. The D&C company built a large mill on the Eureka waterfront, and over the next several decades logged through their timber holdings north of the city.

By the late 1920s the D&C had nearly exhausted its northern timber supplies and started laying plans to move its operations up the Elk River drainage southeast of Eureka. The Bucksport & Elk River Railroad built a branch up the next drainage north of Falk to reach the D&C timber. The D&C eventually purchased the B&ER, and in 1933 extended the line north along the bay to a new log dump a short distance south of the Bayside mill. The D&C trains dumped logs into the bay, where they would be rafted three miles north to the mill. The Pacific Lumber Company bought D&C in 1950, and operations continued under their direction until the timber ran out in 1953.

Two disjointed sections of the D&C log dump remain partially intact fifty-five years after abandonment, some pilings near the shore and then the dump structure itself and associated trestlework out in the bay. Bits and pieces of the B&ER grade are still visible in places going up the Elk River valley, and the D&C's big Camp Carson is now a Boy Scout camp.

WARREN CREEK, CALIFORNIA. Two miles of wooden rails laid to the unusual gauge of 45-1/4 inches on a wharf in Humboldt Bay by a corporate predecessor in 1855 gave the Arcata & Mad River Railroad claim to being the first railroad operating in the Golden state. Several extensions built to serve various sawmills between 1875 and 1884 pushed the rails up the Mad River to the town of North Fork, later renamed Korbel. Except for several idle years during the depths of the Great Depression, the A&MR remained a busy and prosperous shortline, especially after partially converting to standard gauge in 1925.

The A&MR's existence became increasingly dependent on the large Northern Redwood Lumber Company and then Simpson mill in Korbel. The A&MR's end came in 1984, when Simpson closed the railroad in favor of a truck to railcar reload near Arcata. The Eureka Southern Railroad purchased and resurrected the A&MR in 1988, but this revival only lasted until weakened bridges and deteriorating track conditions forced successor North Coast Railroad to end operations in 1993. North Coast salvaged the rails in late 1997.

The Warren Creek trestle is the largest of four wooden trestles on the line. The North Coast Railroad Authority still owns the right-of-way, and there are many in the community that covet the grade as a potential trail.

2

STUFFED AND MOUNTED

Very little captured the American imagination through most of the nineteenth and twentieth centuries like the sights and sounds of a working steam locomotive. The machines at the time represented so much, from the promise of technological advancement to the thrill of speed and motion to the beckoning of new opportunities over the horizon. To the lumber industry from the late 1800s through the middle of the 1900s, the steam locomotive represented an efficient way to extract logs from the forests, and in many ways the development of the logging railroad is what allowed the lumber industry to move off the water and become the primary economic driver in large parts of the country. The growing economic clout of the logging industry in turn prompted manufacturers to either develop themselves or perfect and build specialized equipment designed to meet the challenges of the extreme operating environments in which the industry worked.

The rapidly changing face of technology after World War II very quickly relegated the steam locomotive from a wonder to a relic. By this point the logging railroad itself as an institution was rapidly disappearing as roads improved and the logging industry discovered the substantially cheaper economics of truck logging. Changing logging patterns also helped stack the decks against the logging railroad as selective logging or small patchwork clearcuts replaced the landscape level clearcutting of the previous century. This served to significantly reduce the volume of logs moving from any one geographical area, which heavily favored trucks over trains by itself. Only a few logging railroads survived past the middle 1950s, and those that did typically consisted of little more than a mainline haul from one or more truck to rail reloads down to the mill. The bulk of these also closed by the early 1980s, and the very last logging railroad, operated by Western Forest Products on the north end of Vancouver Island, shut down after suffering a severe accident near the end of 2017. Logs still move by rail on common carriers in a few places, but the book on logging railroads has apparently closed.

The logging railroad was not the only old timber industry equipment or tradition technological advancement relegated to the dust bins of history. The rapid disappearance of almost anything usually triggers preservationist mindsets and movements in concerned circles, and museums, municipalities, and similar institutions started seeking out old logging equipment to preserve and display. Several museums dedicated to only logging industry displays have been created, and many other museums display old logging equipment. The second part of this book profiles logging and lumbering industry related equipment placed on display in various setting around the west, starting with locomotives and moving on to other types of logging relics.

ANGELS CAMP, CALIFORNIA. George Best established the Best Manufacturing Company in an old plow shop in San Leandro, California, in 1884. The Best company built its first steam tractor three years later. Smaller logging and lumber concerns that couldn't justify railroads found the steam tractors useful in both hauling logs out of the woods and then sometimes lumber to the outside world. The tractors also saw extensive use in building logging railroads and roads in later years.

Representative of many steam tractor operators was the Manuel Mill. The Manuel family operated several mills in Calaveras County from 1878 until around 1955. The largest of these lay on San Antonio Creek near the small town of Arnold. The Manuels used horses or oxen to haul logs to the mill and lumber to market until around 1904 or 1905 when they bought two Best traction engines. The engines would haul logs to the mill and usually five loaded lumber carts a day twelve miles from the mill to the railhead at Angels Camp. Trucks eventually replaced the Best machines, but one of them—named Ol' Beth—survived and is on display in front of the Angels Camp museum along with one of the four-wheel carts used to haul logs and lumber.

SONORA, CALIFORNIA. Traditional steam locomotives generally proved themselves unable to cope with the steep grades, sharp curves, and often hastily built and poorly engineered trackage typical of many logging railroads. Michigan logger Ephraim Shay designed and built the prototype of what would become the Shay geared steam locomotive, and he received a patent on the design in 1881. The Shay locomotive featured cylinders mounted vertically on the right side of the boiler that drove a coupled drive shaft extending the length of the locomotive. This drive shaft powered each axle by gears. Shay sold the patent to his locomotive to the Lima Locomotive Works of Lima, Ohio, who went on to build approximately 2,770 Shays in two, three, and four truck varieties between 1878 and 1945.

Lima built this 2-truck Shay as their construction number 2297 in 1910 as the Standard Lumber Company's Sugar Pine Railway #3. Standard Lumber at the time cut logs into lumber at four mills scattered in its holdings in the Sierras, and the Sugar Pine Railway hauled that lumber down to the planing mill, sash and door factory, and other facilities in Sonora. The #3's role would shortly change, as in 1913 Standard replaced its mountain mills with a single large mill in Standard, which eventually grew to include all the Sonora manufacturing facilities as well. The #3's job shifted from hauling lumber to hauling logs down out of the Sierras.

Larger locomotives eventually bumped the #3 out of front-line service, but it continued toiling for Standard and then-corporate successors Pickering Lumber Company and Pickering Lumber Corporation until they retired it in 1957. Pickering placed the #3 on display outside the mill gate in Standard, where it sat until moved to the Sonora fairgrounds in 1977.

CORVALLIS, OREGON. The Baldwin Locomotive Works built the first locomotives with two pilot wheels, eight driving wheels, and two trailing wheels for a Japanese customer in 1897, and for that the locomotive type became known as the Mikado. Locomotives of the Mikado type tended to be well balanced and powerful machines, as the trailing wheels allowed them to have larger fireboxes than those without while also improving the tracking characteristics and ride stability.

Baldwin built its construction number 56399 in May 1922 for the Brooks-Scanlon Lumber Company of Bend, Oregon, as their #5. The factory-installed superheater and piston valves made this one of the more modern Mikados built, but at the same time, its forty-four inch drivers and seventy-ton weight class made it among the smallest. The #5 spent thirty years hauling the cream of central Oregon's pine forests to the mill before diesels replaced it in 1952.

The #5 almost certainly would have gone to scrap when Brooks-Scanlon shut down its logging railroad in 1956 but for a twist of fate. On March 26, 1956, a fire destroyed the machine shop and two locomotives on Georgia Pacific's logging railroad running north from Toledo, Oregon. GP was not yet ready to close the railroad and thus needed immediate replacement power, and some sources indicate Brooks-Scanlon got the #5 back from a Portland scrap yard just in time to sell it. GP retained the number 5 on the machine, and it served that road well until it too closed in 1959. GP subsequently presented the #5 to the City of Corvallis in 1960, and it's been on display in the city's Avery Park ever since.

SNOQUALMIE, WASHINGTON. Substantially built logging railroads hauling lots of tonnage started seeking out larger locomotives from the manufacturers. Baldwin responded with 2-6-6-2 mallet type locomotives in both tender and tank versions, the difference being the tank engines carried water in tanks mounted alongside or slung over the boiler and fuel in a bunker behind the cab instead of both carried in a separate tender.

Baldwin built its construction number 59701 in December 1926 for the Ostrander Railway & Timber Company of Ostrander, Washington, as their #7. The locomotive worked thirteen years there before Ostrander sold it to Weyerhaeuser in 1939 for use on its logging railroad out of Klamath Falls, Oregon, as their #5. Weyerhaeuser removed most of the original saddle tanks in favor of a tender converted from an 8,000-gallon tank car. In 1950 the locomotive passed through the hands of a dealer on its way to a third logging railroad and road number, the #11 of the Kosmos Timber Company of Kosmos, Washington. The #11 closed out its career working for Kosmos and then its successor U.S. Plywood until 1960, when the operation closed.

U.S. Plywood donated the #11 to the University of Washington, who displayed it on their Seattle campus through 1973. The Puget Sound Railway Historical Association moved the #11 to Snoqualmie, where they restored it to operation. The #11 hauled tourists between Snoqualmie and North Bend on and off until 1990. The Association has since become the Northwest Railway Museum, and they completed a cosmetic restoration on the #11 in 2002. It's been placed on display at various places around the museum grounds since.

FLAGSTAFF, ARIZONA. Vast Ponderosa pine forests in the mountains of northern Arizona made the region a major lumber producing area. The Arizona Lumber and Timber Company purchased a mill and commenced operations in Flagstaff in 1887. The AL&T grew, and in late 1910 they placed an order with Baldwin for a 2-8-0 to keep up with their expanding logging railroad. Baldwin's construction number 35938 departed their factory in 1911, wearing the fairly standard olive green and aluminum paint of new Baldwin locomotives and lettered as the #25 of AL&T affiliate Greenlaw Lumber Company. Greenlaw operated a mill in Cliffs, a town just east of Flagtaff that has now been assimilated into the larger community. The #25 worked thirty years for Greenlaw, AL&T, and the Central Arizona Railroad.

In 1941, the Saginaw & Manistee Lumber Company, originally from Michigan, ran out of trees to supply its mill in nearby Williams. The S&M leased and then bought AL&T. S&M reboiled the #25 in 1950, and then in 1953 Southwest Lumber Mills bought S&M. By this time canvas water bags had rubbed the "5" off the cab, and SLM renumbered it to #2. SLM morphed into Southwest Forest Industries in 1960, and the sixty-five-year-old machine was still in use as a backup to the company's diesel when they closed the railroad in 1966. The #25 sat on display in front of the Flagstaff mill until it too closed, at which time title passed eventually to the city, who moved it to a permanent display stand next to the passenger depot downtown.

FLAGSTAFF, ARIZONA. The #25 on the previous page is somewhat unusual in the transitory logging industry for having served its entire working life in one place. The same cannot be said for the other surviving steam locomotive from the same operation. Baldwin built its construction number 60870 in 1929 as Hammond Lumber Company's #6. Hammond initially used the 2-6-6-2T on its logging railroad out of Mill City, Oregon, before transferring it south to Samoa, California, in 1931, where Hammond renumbered it #12. The #12 worked for Hammond's various companies until they bought a diesel in 1950. In 1951, Hammond sold the #12 to the neighboring Northern Redwood Lumber Company, who assigned it to the Arcata & Mad River Railroad after it proved too heavy for their logging railroad. Diesels replaced the #12 again by 1956.

The Saginaw & Manistee and then Southwest Lumber Mills had been using another former Hammond 2-6-6-2T out of Flagstaff since 1944, but it suffered a scorched crown sheet in the middle 1950s, and SWL bought the #12 to replace it. The company removed the saddle tanks after it arrived in Arizona. The #12 was badly worn, however, and in 1959 SWL replaced it with a used Baldwin S-12 diesel locomotive. Southwest Forest Industries subsequently donated the locomotive to Coconino County, and it's been on public display ever since, first as Coconino County Park and now at Arizona Historical Society's Pioneer Museum towards the north end of Flagstaff.

SUTHERLIN, OREGON. The 2-6-2 "Prairie" type wheel arrangement was perhaps the most common one found on logging locomotives, as they had the same riding and tracking abilities as Mikados, but on a short rigid wheelbase that made them adept at navigating sharp curves. Several manufacturers built numerous Prairies for logging railroad service in both tender and tank versions in the same fashion as previously described for Mallets.

American Locomotive Company (Alco) construction number 62965 is typical of the small logging Prairies. The locomotive did not start in the logging industry, however, as Alco initially built it in 1921 as the Hetch Hetchy Railroad #5. Hetch Hetchy's primary role was to support the City of San Francisco's construction of the Hetch Hetchy dam project in the Sierras. Weyerhaeuser Timber Company bought the locomotive in 1935 for use on its extensive logging railroad system in southwest Washington as their #100.

The #100's tenure in Washington lasted until 1948, when Weyerhaeuser broke ground on an entirely new seventeen-mile-long logging railroad running east from Sutherlin, Oregon. Weyerhaeuser equipped the road with the #100, a four-wheel caboose, and 100 log cars transferred south from Washington. The #100 provided all the power the road needed until a diesel arrived in 1956. Weyerhaeuser closed the railroad in 1961 and donated the #100 and the caboose to the city, who have displayed them together in the park ever since.

DUNSMUIR, CALIFORNIA. Only one company, the Willamette Iron and Steel Works, successfully copied the basic Shay locomotive design after the patent expired in 1898. Willamette took advantage of Lima's reluctance to update the Shay by incorporating the latest improvements and technologies into its locomotives, and that, plus their factory location in Portland, Oregon, gave the company a substantial initial advantage over Lima in serving the industry on the West Coast. Lima responded quickly by modernizing the Shay, but not before Willamette built thirty-three locomotives between 1922 and 1929.

Willamette built their construction number 21 in 1926 for the Anderson-Middleton Lumber Company of Cottage Grove, Oregon, as their #2. In 1933, Anderson-Middleston sold the locomotive to Western Lumber Company of Westfir, Oregon, where it worked as the #2 of a succession of owners, including Westfir Lumber Company (1935-1945) and then Edward Hines Lumber Company (1945-1949). Hines sold the machine in 1949 to Medford Corporation, who renumbered it to 7. Medco scrapped out its logging railroad in 1962, but the #7 remained stored in the mill until 1965 when the company conveyed it to a group planning a sizeable railroad museum just south of Dunsmuir. The museum morphed into the Railroad Park Resort, and the locomotive has been welcoming guests onto the grounds since the resort opened.

GARIBALDI, OREGON. Baldwin built its first Mikado in the 90-ton weight class for an Alabama iron concern in 1904, followed by the first of the type employed in the lumber industry in 1907. While typically too big for most log spurs, the 90-ton Mikado offered by several manufacturers quickly became a staple of the logging railroad industry, especially on mainline hauls as logging extended ever deeper into the woods.

Typical of the later Mikados is Baldwin construction number 59071, built in 1926 as Polson Brothers #90. One of the Polson brothers is said to have exclaimed the locomotive would pull every log off the Olympic Peninsula when it first arrived on their railroad north of Grays Harbor, Washington. As it was, the #90 worked faithfully for Polson and then corporate successor Rayonier until replaced by diesels in 1962.

The Oregon Memorial Steam Train Association offered Rayonier $1,980 for the #90, but Rayonier let them have it for $10. The Association moved it to Garibaldi, Oregon, where it has been since displayed along with a passenger car and caboose adjacent to Southern Pacific's Tillamook Branch, later the Port of Tillamook Bay Railroad. The #90 has been repainted and inauthentically lettered for the Southern Pacific since the date of this photo.

IONE, CALIFORNIA. Northeastern California's McCloud River Railroad ordered from Baldwin three similar Prairie type locomotives in late 1900. The first of the trio was McCloud River's #8, Baldwin construction number 18595, built in January 1901. The three helped the road recover from several failed motive power experiments and set the pattern for the modest fleet of Prairie and Mikado type locomotives the railroad would later acquire.

The #8 pulled log and lumber trains through the pine forests east of Mt. Shasta for thirty-five of the next thirty-eight years, with the three-year break coming during the depths of the Depression. Larger power rendered the locomotive surplus, and in 1939 McCloud sold it the Amador Central Railroad of Martel, California. Amador Central renumbered the locomotive to their #7, and it worked another six years hauling lumber down the hill to the Southern Pacific interchange in Ione until replaced by a diesel. The railroad retained the #7, by now tagged with the nickname *Iron Ivan*, as backup power for a few years before awarding it to Ione for display in their city park. The #7 remains there today, though the recreational motorcar groups that now own the remnants of the Amador Central periodically contemplate restoring it to operation.

OROVILLE, CALIFORNIA. In 1920, lumberman R. L. Hutchinson moved his operations from West Virginia to California. The Hutchinson Lumber Company built a sawmill in Oroville and a logging railroad extending from a connection with Western Pacific's mainline north of town into its holdings in the mountains to the east. The venture opened right as lumber markets softened, which forced the company into bankruptcy. Andrew Land, a partner of Hutchinson, bought the remnants and conveyed them to his newly incorporated Feather River Pine Mills (FRPM).

The new company had barely started up when a fire destroyed the Oroville mill. Land let the property sit idle until markets improved in 1938, at which time he built a new mill at Feather Falls and incorporated the Feather River Railway to take over operations of part of the logging railroad. Georgia Pacific bought the operation in 1955, only to shut the railroad down in 1966 after it became apparent the new Oroville Dam would inundate parts of the railroad.

Hutchinson bought three new Shays and one Willamette for the railroad. Lima built the #1 as their construction number 3169 in 1921. The #1 worked forty years for Hutchinson and its successors until 1961, when Georgia Pacific donated it to the City of Oroville. The #1 has been on display in the city's Hewitt Park ever since.

TOLEDO, OREGON. Another locomotive fortunate enough to have remained in one place its entire existence is Baldwin construction number 55347, built in 1922 as Manary Logging Company #1. The Manary story started during the First World War when the U.S. Army Signal Corps entered the spruce logging and lumber industry directly so as to increase spruce production, then required for aircraft construction. After the Armistice, Lumberman C. D. Johnson purchased timberlands in the Siletz, Oregon, area, along with a logging railroad the Army started to build extending north from Toledo, Oregon, towards the timber. Johnson set up the Manary company under the direction of James Manary to handle the logging end of the operation. Manary's involvement ended in 1928, at which point the C. D. Johnson Lumber Company assumed operations.

Georgia Pacific bought the C. D. Johnson company in 1951, which initially changed little but the name on the tender. The #1 was not among the two locomotives burned with the shop on 26 March 1956, which is what brought the former Brooks-Scanlon #5 to Toledo as previously discussed. Georgia-Pacific donated the #1 to the City of Toledo after closing the logging railroad in 1959. The City placed the locomotive on display, and it's now the centerpiece of a collection of railroad and logging relics in the custody of the Yaquina Pacific Railroad Historical Society.

BOULDER CITY, NEVADA. The Pacific Lumber Company started building its mammoth sawmill in Scotia, California, in the middle 1880s. Work included a new railroad extending north towards the docks on Humboldt Bay where lumber would be loaded onto ships, plus an extensive logging railroad system.   The Northwestern Pacific later used substantial parts of the old TPL railroad as part of their mainline connecting the Eureka area with the outside world.

A need for mainline power sent TPL to Baldwin in 1924 for a 90-ton Mikado, construction number 67538. TPL assigned the locomotive their #35, and it settled down to a long career hauling redwood logs down to Scotia. Several TPL log hauls involved trackage rights over the Northwestern Pacific, which required the locomotive to carry Southern Pacific style numberboards. On those hauls, TPL provided the equipment while NWP provided the crews, which became a popular assignment due to the good mechanical condition and cleanliness of TPL's equipment.

Diesels relegated the #35 to storage by the early 1960s. TPL sold the #35 to some railfans, who never moved it from Scotia. In 1971, they resold it to the Heber Creeper Railroad in Utah, who moved it east for use on their tourist railroad. The Nevada State Railroad Museum bought large parts of the Heber Creeper collection, including the #35, in 1992 to equip its new Boulder City campus. The #35 has been cosmetically restored and is now lettered for the museum's Nevada Southern Railway.

SCOTIA, CALIFORNIA. Charles L. Heisler patented the geared steam locomotive bearing his name in 1892. Heisler's design featured two large cylinders inclined in a "V" powering a drive shaft under the boiler's centerline. The shaft powered one axle in each truck, with the side rods transferring power to the other axle. The large cylinders and simple gear arrangement made the Heisler the fastest of all the geared steam locomotives. Heisler assigned his patents to Stearns Manufacturing Company in 1894, who along with corporate successor Heisler Locomotive Works built roughly 625 Heislers in a production run lasting until 1941.

Heisler built its construction number 1446 in March 1921 as Mt. Tamalpais & Muir Woods Railway #9. The #9 hauled tourists to the top of Mt. Tamalpais until plunging revenues forced its sale in 1924 to the Siskiyou Lumber Company of Macdoel, California, who renumbered it #1. The Dolbeer & Carson Lumber Company bought it in 1938, renumbering it to #5. The locomotive worked for D&C until the timber ran out in 1953.

The Pacific Lumber Company had purchased the D&C in 1950, and they retained the #5 for permanent display outside their company museum. TPL officially acquired the #5 on October 5, 1953, renumbered it back to #9, and brought it into Scotia, where it joined a Washington Iron Works Duplex Flyer and a smaller steam donkey. The "permanent display" ended up lasting sixty-five years, as concerns over asbestos and lead paint caused the Scotia Community Services District to erect fences around the display and then auction the equipment off. The Friends of No. 9 submitted the winning bid, and in late November 2018 they trucked the #9 back to Marin County for eventual display somewhere along the old Mt. Tamalpais & Muir Woods grade. The Timber Heritage Association of Eureka got the two donkeys for their help in the move.

KLAMATH FALLS, OREGON. Changing logging techniques and simple economics doomed most logging railroads by the end of the 1940s. Diesel-electric locomotive manufacturers recognized a market for their products existed on those that survived, and several demonstrator units toured these roads. Among them were Baldwin model DS-4-4-750 demonstrators #750 and #751, construction numbers 74814 and 74815, built in June 1950 mostly to promote the forthcoming S-8 and S-12 models. The pair worked on twelve logging railroads or lumber related shortlines in the summer and fall of 1950, four of which went on to buy new S-8s or S-12s, while three more later bought used Baldwin diesels.

Weyerhaeuser bought both demonstrators for use on their logging railroad out of Klamath Falls, followed closely by two S-8s in 1951. The quartet of Baldwins had a career spanning thirty-two years before Weyerhaeuser retired them in 1983. The #101, formerly the #750, is the only survivor of the four. The unit is now owned by the Great Western Railway Museum, and it's been on display in the site of the Oregon, California & Eastern railroad yard in Klamath Falls since the early 1990s. In the background rises the burner of the long-closed Ewauna Box Company sawmill, active from 1912 to 1948, and one-time operator of several logging railroads in the Klamath country.

YACOLT, WASHINGTON. Brooks-Scanlon Lumber Company tested the Baldwin demonstrators #750 and #751 in the summer of 1950, but elected instead to purchase two new Alco model S-3 diesels new in 1952. The pair, Alco construction numbers 79538 and 79774 and numbered B-S #101 and #102, brought logs down into Bend until the company closed their logging railroad in 1956.

The Oregon & Northwestern Railroad then purchased the pair for use on both their railroad mainline between Hines and Seneca as well as the Edward Hines Lumber Company logging railroad east of Seneca. O&NW sold the #101 in 1964 to Longview Fibre Company of Longview, Washington, where it worked for many years as their #7001. The Battle Ground, Yacolt, & Chelatchie Prairie Railroad Association acquired the unit in 2002, and it's pictured here in storage in Yacolt. Longview Fibre added the extension onto the side of the cab reportedly to accommodate a very rotund engineer they employed. The #102 lasted on the ONW until 1968; it subsequently worked for the City of Prineville Railroad of Prineville, Oregon, as their #103, then the Kewash Railroad of Keota, Iowa, and finally the Dakota Southern Railroad of Chamberlin, South Dakota, where it's presently stored out of service. *Jim Heringer photograph*

WILLITS, CALIFORNIA. Two diesels with lumber industry pedigrees share track space in the Roots of Motive Power museum. The green unit is a Baldwin DS-4-4-1000 (SC), construction number 74193, built in 1949 for the U.S. Army Corps of Engineers as their #W8380. The unit passed through the hands of Pan American Engineering Company before landing on the California Western Railroad in 1956 as their #53. The #53 spent the next twenty-eight years hauling trains of redwood logs and lumber on the forty miles of track between Fort Bragg and Willits. California Western sold the #53 to a private party in 1984, who resold it to another party in 1993. Roots of Motive Power acquired the #53 in 1995.

The red unit coupled to the #53 is an American Locomotive Company model S-3, construction number 78399, built in October 1950 as Humble Oil refining #997 at Baytown, Texas. The unit went on to work for Tacoma Municipal Belt in Tacoma, Washington, as their #906 before passing through the hands of a dealer and private party to the Simpson Timber Company of Shelton, Washington. Simpson renumbered the unit #600 and put it to work mostly switching their log dump at Shelton. Roots of Motive Power acquired the #600 from Simpson and trucked it to Willits in October 2005.

CHILOQUIN, OREGON. Sawmills tend to be sprawling facilities, especially if the operation contained substantial yards for drying and stockpiling lumber. Lumber would be handled several times and often would need to be moved substantial distances within a sawmill from the time a log entered the mill to the time the finished product departed the complex.

The most efficient method to move lumber within a sawmill in the days before straddle carriers and forklifts often was a narrow gauge in plant switching railroad, and the vast majority of large sawmills featured such an operation. Fire and pollution concerns ruled out the use of steam locomotives on most of these, and most sawmills employed gravity, animals, humans, or small battery or gasoline powered locomotives. Typical of the battery powered locomotives is this six and a half ton thirty-inch gauge machine, Baldwin construction number 61253, built in March 1930 for the Kesterson Lumber Company of Klamath Falls, Oregon. The locomotive worked for Kesteron and then Klamath Lumber and Box Company before landing in the collection of the logging museum at Collier Memorial State Park in 1984, where it's on display along with a four-wheel flatcar typical of such operations.

MINERAL, WASHINGTON. Early or smaller logging companies that could not afford or justify a logging railroad would often build pole roads instead, which largely consisted of "rails" made of logs laid end to end. Logs would typically be moved on small four wheeled carts, usually powered by animals or specially equipped small locomotives.

One of the more popular locomotives used on pole roads involved mainly Fordson tractor conversions offered by close to thirty different companies across the United States and Canada. Washington's Skagit Steel & Iron Works was one of the principle companies building such conversions on the west coast. Skagit built this machine about 1923 for a Gus Baxtrom of Kosmos, Washington. It survived in several private collections into the area until Jack Anderson brought the machine to the Mt. Rainier Scenic Railroad, now the Mt. Rainier Railroad and Logging Museum.

FLAGSTAFF, ARIZONA. A wheelwright in Manistee, Michigan, named Silas Overpack is generally credited as developing the High Wheels in the early 1870s, though the original idea lay with a farmer who engaged Overpack to build the tall wheels. Overpack's wheels, known by many names including Michigan Wheels, High Wheels, and Big Wheels, consisted of a pair of wheels nine to twelve feet in diameter connected by an axle. When in use, the wheels would be backed over a log, which would be connected to the axle with a chain. Horses, oxen, steam tractors, and then in later years tractors would be used to pull the wheels and the attached logs out of the woods.

Overpack exhibited his wheels at the 1893 World's Columbian Exposition in Chicago. Up to that point the wheels had been a mostly regional phenomenon, restricted to the upper Midwest and the south. The Exposition gave them nationwide exposure, and they very quickly spread to the west, especially in the relatively flat and open pine forests east of the Cascades. Overpack shortly formed a partnership with the Redding Iron Works in Redding, California, to handle the sudden booming demand for the wheels on the west coast. Redding Iron Works made several substantial improvements to the design, perhaps the most important being the "Slip Tongue" model that would lower the log to the ground while going downhill, serving as a brake. This set of Big Wheels welcomed westbound travelers on Route 66 to Flagstaff, Arizona, for many years until restored in the early 2000s and moved to the railroad depot downtown

BEND, OREGON. Newer technologies such as the steel logging arch effectively replaced high wheels by the early 1930s. A good number of preserved examples exist today. Brooks-Scanlon presented this set of wheels seen here to the City of Bend, Oregon, reportedly after the city turned down the Shay locomotive first offered to them. The city has displayed them at their Drake Park, save for some time they were removed for a thorough restoration.

GARIBALDI, OREGON. Steam donkeys are popular display items throughout timber country. Timber companies saved some as mementos to their history, while local museums or historical societies preserved others. Some have been salvaged from where they had been abandoned in the woods for preservation and display. Locations with large numbers of donkeys on display include the Collier State Park, Chiloquin, Oregon; Camp 18 Restaurant and Logging Museum, Elsie, Oregon; and Fort Humboldt State Park, Eureka, California. Perhaps the most unusual location for a displayed donkey is in Disney's California Adventure theme park.

The example seen here is a Washington Iron Works 9x10 machine, construction number 789, built in October 1903. This donkey worked a lot of years for the Silver Lake Railway & Lumber Company in Ostrander, Washington, before going to an unknown mill near Castle Rock, Washington. The Garibaldi Lions Club purchased the machine in August 1969 and added it to their Lumberman's Memorial Park, directly adjacent to the former Polson Brothers/Rayonier #90 previously discussed.

CAMP 20 RECREATION AREA, CALIFORNIA. Perhaps the rarest type of all steam donkeys is the one that started out abandoned in place and then became a display item without moving. When the Caspar Lumber Company shut down its logging railroad in 1946, it dragged this Willamette Iron & Steel Works 10x11 donkey, probably purchased new by the company around 1912, out of the woods to its then abandoning Camp 20. The Jackson State Forest has since developed the site into a roadside rest area, with the donkey placed on prominent display just off Highway 20. Just visible in the trees beyond is the large barn that once held the tractor repair shop at this camp, it has since been demolished.

BEND, OREGON. With the possible exception of the chainsaw, it would be hard to find a single tool or technology that changed the face of the logging industry as rapidly as the "Cat 60" tractor. The C. L. Best Traction Company and Holt Tractor Company merged in 1925 to form the Caterpillar Tractor Company; both manufacturers had previously made 60-horsepower tractors that had been gaining popularity in the logging industry, but the Caterpillar Sixty almost immediately became the gold standard in the woods after the merger. The "Sixty" replaced almost all animal and most steam powered equipment in the woods within a few short years.

Perhaps one of the finest examples of the Caterpillar Sixty on display anywhere is this model on the grounds of the High Desert Museum not far south of Bend, Oregon. Displayed with it is an early version of the steel logging arch that dramatically improved the efficiency and effectiveness of tractor logging. Unfortunately, the industry almost immediately determined these arches inflicted extensive damage to young trees and forest soils.

MINERAL, WASHINGTON. The Caterpillar Sixty reigned supreme in the woods for only a few short years before rapidly evolving technologies quickly eclipsed it. Caterpillar released its first diesel powered tractor into the world in October 1931, and it did not take long for diesel equipment to replace gas equipment in the woods. In 1935, Caterpillar changed its model designations, assigning its diesel tractors RD followed by a number. Top of the new line was the RD-8, which produced 132 horsepower. Caterpillar produced RD—models only from 1935 until 1937, when the company changed the designations again by dropping the "R" and shortening it to the D-series the company still uses today.

The tractor and the Carco logging arch shown here display the technologies the industry adapted to tractor logging to address the destructive tendencies the earlier arches. The combination of the winch on the back of the tractor and the fairlead wheels on the top of the arch allowed the front end of logs to be lifted off the ground during skidding operations. This assembly became standard on all equipment through the rest of the tractor and arch logging period. Murray Pacific and Weyerhaeuser respectively donated the tractor and arch to the Camp 6 Logging Exhibit in Tacoma's Defiance Park in the 1990s. The pair moved to the Logging Museum part of the Mt. Rainier Railroad and Logging Museum after Camp 6 closed in 2011.

TALL TIMBER CAMP, CALIFORNIA. The logging industry apparently first experimented with trucks to move logs around 1913. The technology would take a couple decades to fully mature, but once it did the much cheaper initial construction and equipment costs and lower subsequent operating costs as compared to logging railroads radically changed the way logs were delivered to the mills. This Federal truck dating from around World War I is typical of the first generation of log trucks. It's been resting on display just off Highway 108 at the entrance to a Christian camp.

STEVENSON, WASHINGTON. The extent to which logging truck technology matured in a short time is aptly demonstrated by this beautifully restored 1921 Mack logging truck on display at the Columbia River Gorge Interpretive Center in Stevenson, Washington. Note the chain drive powering the rear axle. Trucks like these and their successors replaced most smaller logging railroads by the end of the 1930s. The Gifford Pinchot National Forest donated this truck to the museum.

CHILOQUIN, OREGON. In 1919, Alfred D. "Cap" Collier opened a small mill at Swan Lake, Oregon, about ten miles northeast of Klamath Falls. The mill depended on trucks for all transportation from the beginning, with two Fageol trucks reported as being in use in the summer of 1920 to haul the 40,000 board feet produced daily to the railhead. Collier became deeply interested in preserving the rich history of the timber industry in the region, and in 1948 he and his brother donated 146 acres and their vast collection of artifacts to the State of Oregon, which has developed it into a state park. The logging museum presently has one of the most extensive collection of logging equipment and tools almost anywhere on the Pacific coast, including this vintage Ford truck Collier's company used later in its operations.

TILLAMOOK, OREGON. The roll of motor trucks as they came into use in the woods quickly expanded to rolls beyond logging. The industry and government agencies shortly developed trucks designed to fight fires. This particular truck is a recreation of a vintage wildland fire truck, consisting of a 1930 Ford Model AA truck with five 55-gallon drums plumbed together mounted in the bed. Oregon Department of Forestry's Southwest District restored the truck and created the replica in 2010 and 2011 as part of the agency's 100th anniversary. It's seen here on display at the Tillamook Forest Center, a visitor's center ODF operates to interpret the ecology and history of the Tillamook State Forest, with a special emphasis placed on the great fires that swept through the forest in the first half of the twentieth century.

CHILOQUIN, OREGON. John R. McGiffert designed and constructed the first log loader that would bear his name about a year after he joined the Clyde Iron Works of Duluth, Minnesota. The McGiffert loader featured a hoisting engine mounted on a raised platform set on a heavy frame, with an A-frame boom attached to one end. When in use the bottom of the frame would sit on the ties outside the rails, allowing empty log cars to pass through the frame underneath the loader. McGifferts could move themselves at a top speed of about four miles per hour using wheels on chain driven axles that folded up against the bottom of the platform while in use.

Clyde built around one thousand McGiffert loaders in a production run lasting through about 1930. The machine became one of the most popular loaders in the woods, especially in the upper Midwest, the South, and the West. All were steam powered save for two gas powered machines built in 1929 for Edward Hines Western Pine Company. Several concerns rebuilt their McGifferts with gas or diesel engines, and Brooks-Scanlon of Bend, Oregon, went so far as to mount a gas-powered shovel on the frame of a McGiffert late in its history.

The two McGifferts on these two pages are among only six or seven of these machines known to survive today. Both are on display at Collier State Park near Chiloquin, Oregon. The loader with the wood cab is an early model, construction number 678, built in 1907 for the McCloud River Railroad as their #1751. The booms on early McGifferts could be raised and lowered buy not swung side to side, and for that reason they were often referred to as "Stiff Boom" McGifferts. McCloud River Railroad sold the #1751

to the McCloud River Lumber Company in 1922, who resold it around 1926 to the Kesterson Lumber Company of Dorris, California. Weyerhaeuser Timber Company bought it in 1929 for use on its Klamath Falls operations, and they donated it to Collier in 1961. Three former Brooks-Scanlon log cars from Bend in badly deteriorated shape are displayed with this loader.

Resting nearby is a machine from the other end of Clyde's production run, construction number 1281, built in 1926 for Big Lakes Box Company as their #42. The machine spent its entire working life in the Klamath Falls area, working for Palmerton Lumber Company and then Weyerhaeuser after Big Lakes. Weyerhaeuser last used it in 1961 and donated it to Collier in 1963. Other known surviving McGifferts are two in the collection of the Southern Forest Heritage Museum in Long Leaf, Louisiana; one beautifully restored at the Lake Superior Railroad Museum in Duluth, Minnesota; and one resting upside down in a lakebed in British Columbia. The possible seventh was reported to be on display in El Salto, Durango, Mexico, as late as 1999 and may still be there.

ELSIE, OREGON. The locomotive crane was hands down one of the most versatile pieces of equipment a railroad logger could own. Cranes could be used for an almost infinite variety of tasks, including cleaning up wrecks, loading logs onto railcars in the woods and/or dumping logs off cars into the mill pond, excavating drainage ditches, dredging log ponds, driving piles for new trestle or bridge construction, moving lumber around within a drying yard, and when needed loading an entire log camp onto flatcars so it could be moved to a new location. The cranes also could move themselves from place to place, albeit slowly.

The market for such cranes was such that nearly ever manufacturer produced a "Loggers Special" aimed at the timber industry. The Ohio Locomotive Crane Company of Bucyrus, Ohio, built this machine for the Long-Bell Lumber Company as their #6. The crane served the company at its operations in Longview, Washington, and then Grand Ronde, Oregon. Long-Bell corporate successor International Paper donated the crane to the Camp 18 Restaurant and Logging Museum.

MINERAL, WASHINGTON. The Clyde Iron Works of Duluth, Minnesota, introduced its "Loggers Special" locomotive crane in 1928. Clyde by this point had a long history of building logging equipment, including the McGiffert loader previously discussed. Unlike most other cranes, Clyde built their Logger Special crane with gas power from the beginning, which cut operating costs and reduced fire danger.

Clyde displayed its first Loggers Special crane, construction number 303, at the 1928 Pacific Logging Congress in Portland, Oregon. George Draham of Olympia, Washington, based Mud Bay Logging Company bought the display unit on the spot. The first has since become one of the last survivors; Weyerhaeuser purchased it for their Vail-McDonald operation in 1941, who in turn donated it to Western Forest Industries Museum. It's now on display on the Mount Rainier Railroad and Logging Museum. The crane is seen here sandwiched between a skeleton log car and an American Car & Foundry 8,000-gallon tank car of a type almost ubiquitously used across the logging railroad industry, also donated to the museum by Weyerhaeuser.

TOLEDO, OREGON. As previously noted, demographic changes in the logging industry found an increasing proportion of married men employed in the woods. Alert logging companies started modifying their living arrangements to accommodate employees with families, including establishing semi-permanent communities to house loggers instead of mobile log camps. This translated into ever increasing distances the companies had to transport loggers to get to and from the job site every day. Most logging companies started building or buying speeders capable of transporting loggers to and from the woods.

Henry Gibson, master mechanic at Weyerhaeuser's shop in Vail, Washington, knew he could build better speeders than those available on the market, and in 1933 he founded the Gibson Manufacturing Company. Gibson erected a factory in Seattle in which his company built approximately 300 speeders ranging in capacity from thirty to 100 men until the company closed following Henry's death in 1953. Approximately fourteen Gibsons still exist, such as this one built new for Coos Bay Lumber Company of Powers, Oregon, as their #6. The speeder worked for Coos Bay successor Georgia-Pacific Corporation before being sold to a succession of private individuals in Washington. The Yaquina Pacific Railroad Historical Society added the speeder to their collection around 1998.

WILLAMINA, OREGON. Common carrier shortline railroads owned by lumber companies often provided the only means of transporting people, mail, and express packages to and from remote sawmill communities in the days before paved highways and widespread vehicle ownership. Very few of these railroads, however, generated enough business to warrant dedicated passenger trains, and several manufacturers developed small railbuses to fill this niche.

Mack Trucks, Inc. built its first railbus in 1903. In 1921, they constructed this small railbus for Oregon's Willamina & Grande Ronde Railroad, owned by the Spaulding-Miami Lumber Company and stretching eight miles from their sawmill in Grande Ronde to a connection with the Southern Pacific in Willamina. The W&GR used the car to haul passengers, mail, and less than carload traffic until 1928, when they sold the car to the Condon, Kinzua & Southern Railroad in eastern Oregon. Kinzua Pine Mills built the CK&S to connect their Kinzua, Oregon, sawmill with the end of a Union Pacific branch line in Codon. The CK&S made extensive use of the car on their twenty-four miles of track until they ended passenger operations in 1951.

CK&S then sold the car to the Trolleyland Electric Railway Museum in Olympia, Washington. The next move brought the car back to Fossil, Oregon, for display at the Wheeler County Courthouse as part of the town's seventy-fifth anniversary celebrations. The car remained in Fossil until the Mt. Hood Railroad bought the car for display in Parkdale, Oregon, under the mistaken belief it had once worked on that railroad. Finally, the Willamina Economic Improvement District paid $10,000 for the car and brought it back to Willamina. The District cosmetically restored the car and have placed it on display near the center of town.

CRESCENT CITY, CALIFORNIA. As noted in the previous section, the Hobbs, Wall & Company built a large mill in Crescent City about 1871. The company built a logging railroad running north of the city, later incorporated as the Crescent City & Smith River Railroad. The CC&SR eventually reached twelve miles to the town of Smith River, on the banks of the Smith River. The railroad built a bridge across the river and prepared to extend the line further north, but a major flood washed out the bridge in the spring of 1890, and the company never rebuilt it. The end of the CC&SR came in 1912 when Hobbs-Wall cut out the last of their timber north of town and started building the Del Norte Southern into their holdings south of the city.

Several local residents discovered this old caboose body in 1987, lying on private land about a thousand feet off the CC&SR grade. The landowner donated the car to the Del Norte County Historical Society. Restoration of the car ended up being a long slow process that stretched out over the next twenty years, and the car finally reached the point where it could be placed on public display in the Del Norte County fairgrounds around 2001 or 2002. It's perhaps the only surviving railcar still in this isolated corner of California.

MINERAL, WASHINGTON. Another logging caboose on display is this car, built by Rayonier, Inc., in their shops at Sieku, Washington. Rayonier made extensive use of the car on their logging railroad running south from Sieku into the forests on the northern end of the Olympic Peninsula until operations ended in 1973. Peter Replinger, a locomotive engineer on the nearby Simpson Timber Company railroad, bought the car. In 1989 he sold the car to the Tacoma Chapter of the National Railway Historical Society, who traded the caboose in 1995 to the Western Forest Industries Museum for a former Union Pacific caboose they owned. The car spent the next sixteen years on display and in occasional operation at the Camp 6 Logging Display in Tacoma, Washington. Western Forest Industries Museum then moved the car to their Mt. Rainier Scenic Railroad after Camp 6 closed in 2011. The car remains on display in the Logging Museum established around the Mt. Rainier Railroad shops.

Coupled to the far end of the caboose is former Rayonier 2-6-2 #45, built by Baldwin in 1906 for Rayonier's predecessor Polson Brothers. The #45 had been displayed in Hoquiam, Washington, after replaced by diesels; it arrived on the Mt. Rainier around 1998. The #45 has moved on since the date of this photo, first to the Oregon Coast Scenic Railroad and then back to Hoquiam for display. On the near side of the caboose is a Skagit speeder that worked many years for Weyerhaeuser out of Klamath Falls, Oregon, before moving to the Mt. Rainier Scenic railroad.

CHILOQUIN, OREGON. Weed Lumber Company purchased this American Hoist & Derrick Company log buncher, AH&D serial number 1397, new in 1926. The versatile tracked buncher could be used in a wide variety of roles in any logging operation. Long-Bell successor International Paper donated the buncher to the Collier Museum in 1970.

CHESTER, CALIFORNIA. Few timber companies preserve and display their history like the Collins Company. T.D. Collins entered the lumber business in Pennsylvania in 1855. The family started expanding its operations almost immediately and reached the west coast by the turn of the century. The company today operates sawmills or forest products related plants in Kane, Pennsylvania; Richwood, West Virginia; Wilsonville, Klamath Falls, and Lakeview in Oregon; and Chester, California.

Collins built the Chester sawmill through their Collins Pine Company. The original mill operated from 1943 until 2001, when the company replaced it with a new mill. What sets the Collins company apart from so many of its competitors is their extremely careful forest management practices that actually increase the volume of timber on their private timberlands from one year to the next and the extensive museum the company has developed in the front yard of the Chester mill. The museum building opened in 2007 and features interactive exhibits about the company's history, forest ecology and management, and related subjects. Scattered around the grounds are over a dozen major pieces of historic logging equipment. Seen in this view are General Electric 70-ton diesel electric locomotive #166, bought new by Collins in 1955 to haul lumber out over their subsidiary Almanor Railroad until it closed in 2004; a Ross straddle lumber carrier built in 1954; and a three-wheeled car bought in 1973 that the company principally used to jump start dead batteries on other equipment during the cold winter months. Other equipment includes several 1930s-1950s vintage log and other trucks, the boat once used to move logs around in the pond, and several tractors and arches.

BEND, OREGON. The High Desert Museum has on its grounds an entire small sawmill on display. Robert Lazinka erected the mill on his homestead near Pilot Rock, Oregon, around 1904. The sawmill operated through the late 1920s or early 1930s. The High Desert Museum acquired the sawmill and moved it adjacent to their recreated farm.

The sawmill is representative of the small mills that once existed almost everywhere trees grow. Mills such as this one typically only produced lumber for local uses and markets. The mill is operational today, though now powered by electricity and compressed air instead of steam. The museum uses lumber the mill cuts during special event weekends scattered throughout each summer in building and maintaining its exhibits.

McCLOUD, CALIFORNIA. The beating heart of almost every sawmill during the steam era was a large steam engine. George Corliss patented the first design for the engine in 1849, and the Corliss Steam Engine Company had a virtual monopoly on the market until the patents expired in 1870. Several manufacturers built Corliss engines in the years thereafter. An external boiler supplied the steam, and the flywheel usually powered by a belt an electric generator and/or belt driven machinery through systems of pulleys and shafts throughout the plant.

The Corliss engine seen here powered one of the large pine mills of the McCloud River Lumber Company from 1903 until 1979. The engine now rests on display behind the Heritage Junction Museum of McCloud, Inc.

BEND, OREGON. Congress created the U.S. Forest Service in 1905 to manage the nation's Forest Reserves, renamed National Forests in 1907. The day to day administration of the national forests in the early yeas fell to forest rangers who mostly lived in the forests in small ranger stations such as the one pictured here. The agency built this station in 1933 in the Inyo National Forest in California, where it existed for the next fifty years. The High Desert Museum acquired the structure and brought it to central Oregon, where it has been restored and placed on display in the museum grounds, complete with displays and exhibits on National Forest management. Volunteers, mostly retired from the Forest Service, staff the station in the busy summer months.

3

REPURPOSED

Not all remnants of the timber industry have either been abandoned or placed on display. A lot of locations where the industry once operated and equipment formerly employed in logging and lumbering have found new careers. Many are the former sawmill or related forest product plant sites that have been redeveloped into shopping centers or housing projects, often leaving no traces of the former industry. The buildings sometimes left behind after a timber related business closed can occasionally be used by other industries, often as light manufacturing or warehousing, but more commonly as equipment or vehicle storage. Old office buildings and other ancillary structures associated with timber enterprise can also be easily converted to non-timber related businesses.

Buildings and other physical facilities are not the only traces of the timber industry that have gone on to other roles. The Rails to Trails movement that has converted thousands of miles of abandoned railroad lines to recreational trails has spread to a few former timber oriented shortlines and a handful of logging railroads. The small size and relative mechanical simplicity of locomotives used in logging or lumber related railroading has made them a staple of the tourist railroad industry, and the longevity attained by many woods locomotives ensured a relatively large number would survive. Several former timber dependent railroads now derive most to all of their existence from tourists, sometimes just as an attraction, while others sell themselves as representing and interpreting the history of the industry. Some tourist railroad operations with few to no historic ties to the industry use former timber industry equipment.

The changing face and nature of the timber industry has deprived a large number of towns of the commercial base that in many cases prompted the creation of the town and/or sustained the economy of an entire region. Communities left in the wake of the departure of the timber industry either face immediate abandonment or a long, slow decline. Many former timber towns struggle to find new sources of revenue, with tourism usually being the most prominent focus of economic redevelopment efforts. The third part of this book will focus on profiles of buildings, communities, equipment, and the like in a post timber industry world.

BEND, OREGON. The Shevlin-Hixon Lumber Company placed its new sawmill in Bend, Oregon, into operation on March 3, 1916. Seven weeks later, on April 22, the Brooks-Scanlon Lumber Company opened their mill almost directly across the Deschutes River from the Shevlin-Hixon mill. Both companies had long roots in the timber industry in the Upper Midwest and spread their operations to the south and west. The arrival of two large well financed timber enterprises transformed Bend almost overnight from a sleepy town of less than a thousand people to one of the major centers of the pine lumber industry in the west.

The two mills ran full bore until 1950, when concerns over dwindling timber supplies caused Shevlin-Hixon to sell their interests in the area to Brooks-Scanlon. The Shevlin-Hixon mill closed almost immediately, with the mill buildings demolished in the years thereafter. The slowdown in logging allowed the Brooks-Scanlon mill to remain in operation until 1994, though under the Diamond International Corporation name after 1980.

Bend boomed in the years since the timber industry left and has become a thriving mecca for outdoor recreation in all seasons, partly due to the development efforts of Brooks Resources, the successor corporation to Brooks-Scanlon, and also partly due to a huge influx of retirees. William Smith Properties purchased the 270 acres containing both mill sites after the Diamond International mill closed and has since redeveloped the entire area into the "Old Mill District," an upscale shopping and entertainment center.

Building the Old Mill District required demolishing all but a very small number of the original mill buildings, though several other historic structures have since been moved onto the site. The largest original building still standing is the Books-Scanlon Lumber Company Mill "A," built in 1916 and closed in 1983. The restored structure now houses Tumalo Creek Kayak & Canoes. The concrete blocks in front of the building once supported the power house and sawdust burners associated with the mill. The old Brooks-Scanlon box factory stands not far away, though it is not officially part of the Old Mill District.

A good part of the Shevlin-Hixon mill site across the river from Mill "A" now houses the Les Schwab Amphitheatre, an 8,000-person-capacity performance stage named after Les Schwab, founder of the tire chain bearing his name. While the Les Schwab headquarters are located in nearby Prineville, Les himself grew up mostly in Brooks-Scanlon log camps and often joked he attended "Brooks-Scanlon University" when asked about his college experience. The four camp cars are historic Brooks-Scanlon cars, moved from a former logging camp near Sisters, Oregon, to behind the amphitheater for use as dressing rooms and offices. Also relocated to this site and just visible past the end of the cars is the former Brooks-Scanlon mill office building, now a green room for performing artists. The "Art Station" is in the original Bend passenger depot, built by the Oregon Trunk Railroad when it arrived in town in 1911 and moved to this spot in 1999 to allow for Highway 97 to be widened through town.

BEND CONTINUED. The defining landmark in the Old Mill District are the three towering smokestacks rising from the roof of an old powerhouse. Brooks-Scanlon erected the powerhouse in 1923 as part of their Mill "B" complex built a third of a mile upstream from Mill "A." Mill "B" and its associated facilities remained in operation until 1994.

The old powerhouse building today houses the Bend store of outdoor equipment retailing giant Recreational Equipment, Inc., better known as simply REI. The REI store includes both the former power house and an adjacent fuel storage building, while another old fuel building housing other businesses stands just behind the store. While a lot of the buildings around the REI store are architecturally styled after the sawmill buildings, they are all new construction. The Old Mill District's logo and almost all their promotional material features the three stacks.

Prior to 1950, log trains of both companies dumped their logs directly into the river. A log boom down the middle of the river kept the logs of the two operations apart. You'd never know any of that by looking at the scene now. The Old Mill District invested heavily in removing a lot of the alterations to the river banks the sawmills wrought and establishing riparian vegetation along the banks. Wide paved hiking and biking trails run along both sides of the river, with several pedestrian bridges crossing the water. During the warmer months an endless stream of rafts, canoes, kayaks, boats, intertubes, and like craft drift lazily down the river through the District. Large crowds of people are drawn to the District's many shops, restaurants, movie theatres, and other retail and service businesses. A plate underneath the middle smokestack in the REI store, a handful of interpretive panels scattered around, and a few other remnants of the timber industry displayed around the District are some of the few reminders of what once existed here, but few people take note of them.

FAIRHAVEN, CALIFORNIA. Plywood was first introduced into the United States around 1865, but it did not come into its own as a building material of choice until the immediate post-World War Two era. Soaring demand for the material prompted many established forest product companies to add plywood plants to existing mills, while others built entirely new plants.

Among the latter was the Mutual Plywood Company, which in the winter of 1948/1949 built a new large plywood plant in Fairhaven, California, just down the Samoa Peninsula from Samoa. U.S. Plywood Company purchased the plant in the very early 1960s, only to sell it to Simpson Timber Company in 1965 as part of a series of mutually beneficial transactions between the two companies. Simpson ran the plant until 1981, when timber shortages and market conditions forced its closure.

The old plywood plant is now part of the Fairhaven Business Park, owned since 2005 by Sequoia Investments X, LLC. A year or two after that transaction, FoxFarm Soils & Fertilizer Company, a leading producer and supplier of organic fertilizers, soil amendments, and other products, set up their storage, manufacturing, and distribution operation out of the old Mutual facility. A couple old rail spurs still lead into the plant, though the present occupant has no use for them, and even if they did the old Northwestern Pacific mainline south of Eureka has been impassible due to flood damage since the first days of 1998 anyway.

FALK, CALIFORNIA. The Elk River Mill & Lumber Company engine house was the only substantial building in the old sawmill town of Falk to survive the demolition of the rest of the town described in part 1 of this book. In 2008, the Bureau of Land Management carefully disassembled the badly deteriorated building and the adjacent sand dryer from their original site. The BLM then reconstructed the shop and built a recreated sand house around the dryer at a new site across the river, downstream from their original location, and adjacent to the hiking trail the agency was then developing into the town site and the Headwaters Forest beyond. The reconstruction work incorporated around forty percent of the original materials from the engine house, as the rest had become too decayed to be reused. Iron straps on the floor shows where the rails would have been in the original structure. The buildings now house the "Headwaters Education Center" containing various interpretive panels explaining the natural and human history of the area and how the two interrelate.

HINES, OREGON. The timber industry in general and the Edward Hines Lumber Company in particular demonstrated the longevity and durability of lumber as a building material by profiling Colonial-era wood structures in their advertising, especially in the middle decades of the 1900s when other building materials started making inroads against lumber in new construction. George Washington's Mount Vernon Estate, started in 1742, was one of the most commonly featured buildings in these efforts, and Hines was one of several companies that took this concept to the next level by pattering their office building after the Mount Vernon estate's main structure. The Owen-Oregon Lumber Company introduced this concept to Oregon in 1927 with their new office building for their Medford, Oregon, mill. Two years later, the Edward Hines Western Pine Company erected a near carbon copy of the Owen-Oregon structure for their Hines mill. This building served as the mill office for Edward Hines Western Pine, Edward Hines Lumber Company, and finally Snow Mountain Pine before the mill closed in 1995.

After the sawmill closed the Eastern Oregon Academy, a residential treatment facility for troubled youth, bought the building and set up operations. The Academy operated until the state revoked its license in late 2016 or early 2017. As of this writing the building sits vacant, though it has recently been sold.

WESTFIR, OREGON. The Western Lumber Company built a sawmill and the associated company town in Westfir in 1923. A large U.S. Forest Service timber sale up the Middle Fork of the Willamette River and a contract to log trees on the route of the Southern Pacific's mainline across the Cascades gave the mill a secure resource base from which to work. Westfir Lumber Company acquired the property in 1935, who in turn sold to Edward Hines Lumber Company in 1945. Hines added a plywood plant and made many other improvements to the operations over the next thirty years before selling Westfir in 1977. The sawmill closed immediately, but a quick succession of several owners kept the plywood plant going until a fire destroyed it in 1985.

The timber industry was largely responsible for the economies of the towns of Westfir and nearby Oakridge, and both are now among countless former timber towns in Oregon and Washington trying to figure out what comes next in a post-timber world. The two communities are focusing their efforts on trying to become a mecca for mountain biking and similar pursuits. The former sawmill office building in Wesfir is now a guest lodge catering heavily to the biking market. Hines built the covered bridge over the river in 1945 immediately after buying the town, and today it leads to a parking area where the sawmill once stood that is now the trailhead to the network of biking trails radiating out from the town.

McCLOUD, CALIFORNIA. Friday George built the first sawmill at what he called Sugar Pine Park in 1892, only to go broke two years later. A group of San Francisco based businessmen led by George W. Scott and William M. Van Arsdale bought George's property out of bankruptcy in 1896 and incorporated it as the McCloud River Lumber Company. The Scott & Van Arsdale group replaced George's mill with a much more substantial sawmill, renamed the town built around the mill to Vandale and then McCloud, and built the McCloud River Railroad to both connect the mill with the Southern Pacific mainline and to bring logs from the woods east of McCloud into the mill.

The Shevlin, Hixon, Carpenter, and Clark families, all with long ties to the Minnesota timber industry, bought the McCloud River operations in 1902. The stability, financial backing, and long marketing reach of parent Shevlin Pine Sales resulted in a prolonged period of prosperity for the McCloud River operations. The McCloud mill was one of the largest single pine mills operating anywhere on the Pacific coast, and lumber from its saws spread to all corners of the world. U.S. Plywood Corporation bought the McCloud River Lumber Company in 1964, and then its corporate successor Champion International closed the mill down in 1979. The end had not yet come, however, as P&M Cedar Products purchased and reopened part of the mill, primarily to cut cedar shingles and wooden pencil stock. P&M's corporate successor Cal Cedar closed the McCloud mill probably for the last time in 2003.

McCloud lies in the shadow of California's Mount Shasta, and as the timber industry died the town reinvented itself as destination for the legions of tourists that flock to the area around the mountain. Several operators opened bed & breakfast or similar upscale lodging businesses in some of the old buildings around town. The community became a center for square dancing and built a modest tourist business around that pastime. The railroad, by now operating as the McCloud Railway Company, got into the act with the Shasta Sunset Dinner Train that operated from 1996 until the early days of 2010.

Centerpiece of almost every company town was the company store, through which employees and their families were generally required to purchase all of their needs and desires as a term of employment. The large McCloud store building speaks to its role as the centerpiece of a once vast commercial empire that encompassed not only the town but also satellite operations in the remote logging camps, such as the store in Pondosa featured in Part 1. At one time the company operated its own farms, ranches, dairies, and other facilities to supply the store, and anything the company or its employees needed that wasn't in stock could be ordered in. The McCloud River Mercantile Company now operates the store building, which includes lodging and a restaurant in addition to a store and several other businesses. Just up the hill is the McCloud River Inn, one of the many bed & breakfast operations in the town, located in the former lumber company office building.

SCOTIA, CALIFORNIA. Logging has always been one of the most dangerous industrial occupations in the United States. The frequency and severity of injuries suffered on the job coupled with the remote location of most sawmill towns from major population centers caused most of them to have larger hospital facilities than what the town size might otherwise suggest.

The old Pacific Lumber Company hospital building in Scotia is typical of many such company operated hospitals. PLCo operated the hospital up until the 1950s, when improving highways made it feasible to start sending patients to the larger hospitals in Eureka instead of operating their own facility. The company used the building for offices, archive storage, and other purposes for several decades afterwards. In 2012, the Southern Trinity Health Services Inc. opened a satellite clinic facility in the old building.

PONDOSA, OREGON. The fire that destroyed the by-then closed sawmill and other facilities in Pondosa on June 20, 1959, as described in Part 1, spared the small business district, which largely consisted of a store and hotel. The store building has long since been razed, but some entrepreneurs established the Pondosa Store in the old hotel building.

The once busy sawmill town has now been reduced to the store, a few houses, a couple signs on the state highway, and about a dozen residents. The store remains open, stocking a few basic supplies for the scattered residences and ranches in the greater area and for the bicyclists, mushroom pickers, and other tourists that pass through the place. The store also hosts an annual reunion for the increasingly fewer people who once called Pondosa home.

STARKEY, OREGON. Wisconsin lumberman August J. Stange incorporated the Mt. Emily Lumber Company in 1924. The new company built a sawmill in La Grande, Oregon, together with a logging railroad running from a connection with the Union Pacific mainline at Hilgard, Oregon, west up the Grande Ronde River. The logging railroad eventually extended approximately forty-two miles up the river, and for the next thirty years the Mt. Emily log trains brought the loaded log cars down to Hilgard. Union Pacific moved the log cars the eight miles to and from La Grande.

Like most logging companies, Mt. Emily housed its loggers in a semi-mobile log camp that would be shifted around as needed. The company moved the camp to its final location in 1930. In 1956, a year after the logging railroad shut down, the company donated the camp to the Blue Mountain Conservative Baptist Association, who converted it to their Camp Elkanah. The bulk of the camp today is still in former Mt. Emily Lumber camp cars, most of which are still on wheels on rails, and skid shacks. Most of the cars are still basically as Mt. Emily left them, but they are now seasonally occupied by church groups, Christian summer camps, and other organizations that lease the facilities. *John Henderson photo, Jeff Moore collection*

IMBLER, OREGON. The Mt. Emily Lumber Company principally used a pair of geared steam locomotives to power its log trains, one a Willamette and the other a Shay. When news coverage of the last log trains the company operated in June 1955 trickled back to Portland, the Oregon Museum of Science and Industry became seriously interested in acquiring the Willamette and bringing it back to the city where it had been built for preservation. Mt. Emily management agreed to donate the locomotive, but neglected to inform the scrapper then junking the railroad until after they cut the Willamette up. The museum received instead the Shay, which they subsequently conveyed to the Oregon Historical Society.

The Shay bounced around Portland for a decade and a half before being leased to the Cass Scenic Railroad in West Virginia until 1994, when the lease ended and it returned to Oregon. It's been in occasional operation on the City of Prineville Railroad since. As for the Willamette, the only remnant of it is the water tank, long in use as storage tank for a well on a ranch outside Imbler. *Tom Moungovan photo*

KLAMATH COUNTY, OREGON. Railroad promoter Robert Strahorn laid ambitious plans to build a 400-mile network of lines that would have tied together several railheads in central and eastern Oregon. Strahorn's Oregon, California & Eastern Railroad broke ground in Klamath Falls, Oregon, in 1917, but only managed to build forty miles of track over the next six years. The completed OC&E trackage passed through some prime Ponderosa pine forests, and the OC&E generated a lot of log traffic destined for the big mills in Klamath Falls.

Strahorn sold the OC&E to the Southern Pacific Railroad in 1925, who in turn conveyed a half interest in the company to the Great Northern Railroad in 1928. The new owners extended the mainline 26.4 miles east to the small town of Bly, the only one of several planned extensions completed. The OC&E ceased independent operations in 1933, with SP and GN taking turns operating the line thereafter.

The OC&E survived mostly on log traffic interchanged from the six different private logging railroads with which the line connected. In 1940, Weyerhaeuser Timber Company started building another logging railroad running from Sycan, several miles west of Bly, north into its substantial holdings. Weyerhaeuser gradually became the OC&E's only shipper of any consequence, especially after it bought and modernized the small sawmill in Bly. The interdependency eventually led Weyerhaeuser to acquire the OC&E in 1975.

The OC&E boomed over the next fifteen years as Weyerhaeuser dramatically increased its harvest rates as part of a planned change in forest management practices. However, operations started to slow in the later 1980s as Weyerhaeuser started running out of trees to cut. The axe finally fell on the OC&E and the Weyerhaeuser logging railroad in 1990, and scrappers tore both railroads out in 1992 and 1993.

A local rails to trails group started efforts to preserve the grade of the two railroads as a recreational trail. Oregon State Parks agreed to buy the right-of-way and convert it to the 109-mile-long OC&E Woods Line State Trail. An OC&E caboose is on display at the trailhead in Klamath Falls, and the state has scattered informational kiosks patterned after small railcars at key points on the line.

ARCATA BOTTOMS, CALIFORNIA. The Dolbeer & Carson Lumber Company initially depended on the common carrier Oregon & Eureka Railroad to haul its logs from the area around McKinleyville south to its Eureka mill. The O&E was among the properties A. B. Hammond acquired when he entered the redwood industry, and William Carson's resentment towards Hammond prompted the D&C to build its Humboldt Northern Railroad in 1904 and 1905 so that it could stop using the O&E. The D&C hauled its own logs south over the HN until 1930 when their timber ran out and they shifted their operations south of Eureka.

D&C sold the HN to the Little River Redwood Company, which had been the line's other major shipper. In a twist of irony, the Little River company promptly ran into financial difficulties, and Hammond Lumber purchased it in 1931. Hammond almost immediately shifted its log trains over to the old Humboldt Northern, and they continued using the line until successor Georgia Pacific closed the logging railroad in 1961.

The HN originally crossed the Mad River on a covered bridge, which lasted until Hammond replaced it with this steel bridge in 1941. The bridge now sits at the south end of a five mile stretch of the old grade that has been converted to the Hammond Trail, operated by the Humboldt County Parks & Trails Department. The bridge's days may be numbered, though, as the structure is rapidly corroding and rusting away, and the organizations and agencies associated with the trail are exploring options to replace it.

EUREKA, CALIFORNIA. The lumber industry provided vast financial rewards to a lucky few entrepreneurs, many of whom tended to express their wealth with mansions. Few such estates on the west coast represented this better than the Carson Mansion, built by William Carson of the Dolbeer & Carson Lumber Company. Carson invested $80,000 in building the mansion between 1884 and 1886, and the building is well noted as containing a mix of every major style of Victorian-era architecture. Carson is quoted as saying of his mansion, "If I build it poorly, they would say that I (am or was) a damned miser; if I build it expensively, they will say I'm a show off; guess I'll just build it to suit myself." Noted San Francisco architects Joseph and Samuel Newson designed the building, and specialty lumber from Central America, East India, the Philippines, and Mexico supplemented the local native redwood.

Carson built his mansion on a hill directly adjacent to the D&C mill, and from it he could monitor just about anything happening in his mill. The Carson family occupied the structure until they sold the company to the Pacific Lumber Company in 1950. A group of prominent Eureka businessmen bought the mansion from the family, and they set up the exclusive and private Ingomar Club, named after a theater Carson owned, to enjoy and maintain the structure and surrounding grounds. The D&C mill below the mansion is now a weedy field with only a few concrete blocks hinting of the industry that created the wealth that built the mansion. The mansion itself is closed to the public, but that doesn't stop it from being one of the most photographed buildings in California.

CHILOQUIN, OREGON. Little brother to the "Cat Sixty" detailed in the previous section was the "Cat Thirty," the thirty-horsepower tractor. While it wasn't nearly as large or powerful as the Sixty, the Thirty—nicknamed "Kittens" by at least a few operations—never the less proved their worth as versatile machines in the logging industry. This beautifully restored model is seen here pulling a vintage road grader on the grounds of the Collier State Park during their annual living history event.

GARIBALDI, OREGON. The Saginaw Timber Company was another Michigan refugee that brought their name with them when they moved to Brooklyn, Washington, after exhausting the timber supplies in the upper Midwest. In 1912, Saginaw purchased new from Baldwin their construction number 38967, a saturated steam slide-valve equipped 70-ton Mikado. Saginaw assigned it #2 on their roster.

The #2 settled down to a long career of hauling logs for Saginaw, then North Western Lumber Company, and finally Polson Brothers, all in western Washington. Polson sold out to Rayonier in 1948, but that transaction changed only the name on the side of the tender. Rayonier finally purchased diesel electric locomotives in 1962, which finally brought the #2's fifty year career in the woods to a close.

The #2 was not yet done, as Rayonier subsequently sold it to the Michigan's Grand Transverse Northern Corporation, partly due to mistaken beliefs the locomotive had once worked in Michigan. Subsequent sales took the #2 to the Cadillac & Lake City of Lake City, Michigan; Kettle Moraine Railway of North Lake, Wisconsin; Illinois Railroad Museum of Union, Illinois; and finally, the Mid-Continent Railway Museum of North Freedom, Wisconsin. A contract dispute with Mid-Continent forced the #2's owner to move it again, and the Oregon Coast Scenic Railroad presented the best operational opportunities. OCSR trucked the #2 back across the country in the fall of 2017, and it's now joined that operation's impressive and growing collection of former logging industry locomotives. The move brought it back in contact with one of its Polson/Rayonier stablemates, the #90 featured in Part 2 of this book, as it is on display immediately adjacent to OCSR's small shop and depot area.

ELBE, WASHINGTON. When Polson Brothers Logging sought new locomotives in the early 1920s to replace its older power, their master mechanic recommended they buy a duplicate of the Saginaw Timber 2. The result was Baldwin's construction number 55355, another saturated steam and slide valve equipped Mikado, built in April 1922. The #70 put in forty years on the Polson and then Rayonier railroad, though bumped to switching and work train service by larger power in the later years.

The diesels that forced the retirement of the #2 and #90 in 1962 similarly affected the #70. A railfan named Maynard Lang bought the #70 in 1963. Lang had the #70 stored in Auburn, Washington, until 1965, when he moved it to the Puget Sound & Snoqualmie Falls Railroad in Snoqualmie, Washington. Tom Murray bought the #70 following Lang's death in 1992 and donated it to the Mt. Rainier Scenic Railroad, who restored it to service in 2011. The #70, seen here getting ready for a rainy day of hauling tourists, is still in service on the Mt. Rainier Railroad and Logging Museum.

BARVIEW, OREGON. California's McCloud River Railroad purchased six large new Prairie-type locomotives, two from Baldwin in 1924 and four from Alco in 1925 as part of a program to modernize its locomotive fleet. The last and largest of these was Alco construction number 66435, built as McCloud River #25. It along with its identical twin #24 were the last steam locomotives McCloud purchased new.

The #25 was destined to spend most of the next thirty years primarily leased out to the McCloud River Lumber Company for use on their logging railroad system, with most of that time spent in the camps of White Horse, Widow Valley, and Kinyon. The engine also filled in as needed on McCloud River's mainline freights. Diesels and log trucks replaced the #25 in 1955.

The railroad stored the #25 instead of scrapping it. McCloud River returned the #25 to operation in 1962, which launched a second career hauling tourists on its home rails that lasted off and on until 2008. Tough economic times finally forced McCloud River successor McCloud Railway to sell the locomotive, and in 2011 the Oregon Coast Scenic Railroad bought it. The #25 is now OCSR's principle locomotive on its tourist runs between Garibaldi and Rockaway Beach, with occasional longer runs and special charters operated throughout each year.

FELTON, CALIFORNIA. Lima built its narrow-gauge Shay construction number 2465 in 1911 for a locomotive dealer in San Francisco. The locomotive initially worked as the #4 for the Truckee Lumber Company's Butte & Plumas Railway and then the Swayne Lumber Company, both of Oroville, California. Swayne sold the locomotive around 1917 to the West Side Lumber Company of Tuolumne, where it hauled logs down out of the Sierras as their #7 until that railroad closed in 1960. West Side donated the #7 to the City of Sonora, who placed it on display at the fairgrounds. In 1978, the West Side & Cherry Valley Railroad, a tourist operation on parts of the old West Side railroad, bought the #7, and the Sugar Pine Railway #3 discussed in Part 2 replaced the #7 at the fairgrounds.

The inflation and energy crisis of the late 1970s doomed the West Side & Cherry Valley, and in 1986 the Roaring Camp & Big Trees purchased the #7. The RC&BT has always been a tourist railroad, built in the early 1960s through one of the few remaining privately owned stands of old growth redwoods in the Santa Cruz mountains, but it's always had the deliberate look and feel of a logging railroad. The #7 debuted on the RC&BT in 1997 after a lengthy restoration, and it's been in regular use on that operation since. This picture is of the "boring" side of the Shay on a foggy morning as it prepared for another day's worth of hauling tourists up the mountain.

GOLD HILL, NEVADA. The sparse deserts of western Nevada might seem like a strange place to find a logging locomotive, but the area is home to several. Baldwin built its construction number 44235, a 2-8-0 Consolidation type, new for Long-Bell Lumber Company's Louisiana and Pacific Railway as their #252. Long-Bell transferred the locomotive to Longview, Washington, in 1922, where it became the Longview, Portland & Northern #680. One last move brought the #680 to Grande Ronde, Oregon, where it closed out its logging and lumbering career, first as Willamina & Grande Ronde #680 and then again as LP&N #680 after Long-Bell merged the W&GR into that road.

Diesels replaced the #680, but it remained stored in the Grande Ronde engine house until the Virginia & Truckee Railroad bought it in 1977. Bob Gray and his family had been actively rebuilding a few miles of the fabled V&T out of Virginia City as a tourist railroad and reached the point where he needed his own equipment, and the #680 fit the need perfectly. Renumbered V&T #29 and now named Bob Gray, the locomotive is in regular operation hauling tourists through the old Comstock mining district in the summer months.

MOUND HOUSE, NEVADA. The McCloud River Railroad purchased the first 90-ton logging Mikados used in the timber industry new from Baldwin in 1907. Seven years later, the company bought a third such machine from Baldwin, their construction number 41709. Baldwin records show an Arkansas logging company placed the initial order, but McCloud River assumed it after the original deal fell through and assigned it #18 on its roster. McCloud River displayed the #18 at the Panama-Pacific Exposition in San Francisco throughout 1915 before bringing it home.

The #18 spent the next forty years hauling log and lumber trains through the California pines before being replaced by diesels. In 1956 McCloud sold the #18 to the nearby Yreka Western, where it continued hauling lumber and tourist trains until it blew a cylinder head while hauling an excursion train in 1964. The #18 languished in Yreka until the McCloud Railway returned the locomotive to McCloud in 1998. McCloud Railway restored the #18 to service in 2001, but economics forced its sale in early 2005.

One of the most ambitious projects in the railroad preservation world since the 1990s is the State of Nevada's reconstruction of the Virginia & Truckee Railroad from Gold Hill to Eastgate, just west of Mound House. The Commission managing the project purchased the #18 and shipped it to Nevada in 2009. The Gray's V&T leases the #18 and operates tourist trains over the state's railroad. The #18 has been down for required boiler recertification and other work since the end of 2015, but efforts are on-going to get it back in service.

CARSON CITY, NEVADA. The fabulous Comstock Mining District in western Nevada consumed enormous quantities of wood for timbering mine shafts, building the massive mill and other structures, and in fuel among other uses. The nearby Lake Tahoe basin contained vast pine forests, and in 1873 the Carson & Tahoe Lumber & Fluming Company set up one of the few lumbering operations in the Silver state. The C&TL&F built a sawmill at Glenbrook, Nevada, and started logging around the lake. In 1875 the concern purchased two narrow gauge 2-6-0 "Mogul" type locomotives new from Baldwin, named the *Glenbrook* and *Tahoe*.

Baldwin built the *Glenbrook* as their construction number 3712, and up until the timber ran out and the sawmills closed in 1898 it and the other C&TL&F power moved logs to the mill and then lumber up to the rim of Lake Tahoe, where it would be sent down flumes to the floor of the Carson Valley. In 1898, the Lake Tahoe Railway & Transportation Company purchased the railroad assets of the C&TL&F for reuse in a new railroad line built to bring tourists from Truckee to the lake. Southern Pacific leased the LTR&T and converted it to standard gauge in 1925, which finally put the Glenbrook out of work.

The subsequent history of the *Glenbrook* included long periods of storage and display at various venues in California and Nevada interspersed with a few years spent as a parts source on the Nevada County Narrow Gauge railroad. The *Glenbrook* finally landed at the Nevada State Railroad Museum in 1981, and their restoration shop returned the locomotive to service in 2015 after a lengthy and meticulous restoration.

ELBE, WASHINGTON. Among the relatively small number of traditional rod-driven steam locomotives designed specifically for logging railroad service was Alco's 2-8-2 tank-type Mikados. Alco built twenty-one of these machines for mostly western logging roads. Representative of the class is Alco's construction number 68057, built in 1929 for the Crossett Western Company of Wauna, Oregon, as their #11. The locomotive hauled logs out of the northwestern Oregon forest until 1942, when Crossett sold it to California's Hammond Lumber Company. Hammond renumbered it to #17 but only got three years' worth of use out of it before a massive fire burned out several trestles, stranding the locomotive at a log camp named The Gap. The #17 sat in isolation until Gus Peterson bought it in 1965 and trucked it out to his Klamath & Hoppow Valley Railroad in Klamath, California, where it hauled tourists until high gas prices forced the operation to shut down.

Tom Murray purchased the K&HV equipment in 1980 and moved the pieces to the new Mount Rainier Scenic Railroad. The MRSR shops restored the #17 to operation in 1995, then rebuilt it again between 2011 and 2013. It's been the primary locomotive used on the Mt. Rainier Railroad and Logging Museum trains since.

EUREKA, CALIFORNIA. The economic frenzy of gold mining in California spawned the creation of several iron works and equipment manufacturers in the San Francisco bay area. Marshutz & Cantrell is typical of many such operations that converted over to building equipment for other industries, mostly logging and agriculture, after the gold boom faded. The company built a large number of steam donkeys and around a couple dozen small locomotives, most of them for use in the timber industry on the west coast.

Marshutz & Cantrell in 1884 built a small 0-4-0 locomotive for Noah Falk. Falk used the small machine in his sawmills in Arcata, California, before conveying the machine to the Elk River Mill & Lumber Company as their #1. The #1, named *Falk*, almost single handedly kept the company rolling for almost twenty years. The *Falk* was a direct drive locomotive, with the capstan drum on the front powered by a small third cylinder. A larger locomotive bumped the *Falk* to switching the mill until 1927, when a third locomotive caused its retirement.

The City of Eureka acquired the *Falk* in 1937, and then later conveyed it to the State of California who placed it on display at Fort Humboldt State Historic Park. In 1986, the forerunner of today's Timber Heritage Association restored the *Falk* to service, and it's been in service on a short stretch of track at Fort Humboldt since. The locomotive has also travelled as far afield as Canada to participate in special events through the years.

EUREKA, CALIFORNIA. In 1892, the Bear Harbor Lumber Company acquired a gear driven 0-4-0 from Marshutz & Cantrell that became their #1. Bear Harbor got its start cutting railroad ties and harvesting tan bark, and the #1 initially made its living hauling those products down to the doghole port at Bear Harbor. The company's plans involved pushing the railroad inland to the Eel River, where a sawmill would be built. Constructing the railroad turned out to be a slow process, affected by fires, financial troubles, and one big storm that destroyed the wharf.

The railroad reached the banks of the Eel at the sawmill site in 1905. However, late that year an accident in the sawmill, then nearing completion, claimed the life of one of the principle company managers. His death and then the San Francisco earthquake in 1906 doomed the project, and the sawmill never opened, though smaller mills later occupied the site.

The Bear Harbor railroad lay abandoned for decades, and parts of it may still exist. The #1, by now named *Gypsy*, sat abandoned in its small engine house near Moody until the company moved it to Andersonia, across the river from Piercy, in 1957. The family donated the locomotive to the State of California in 1967, who placed it on display at Fort Humboldt. Timber Heritage Association's predecessor restored it to operation in 1979.

SUMPTER, OREGON. Sumpter is perhaps best known for its gold mining history and heritage, but timber is what brought the town its railroad connection. The Sumpter Valley Railroad, organized by the backers of the Oregon Lumber Company, started pushing its narrow-gauge rails west from Baker City in 1890. The railhead reached Sumpter in 1896 and remained there for four years before the SVRy pushed its rails to the southwest, eventually reaching Prairie City, Oregon, in 1910. The Sumpter Valley hauled primarily logs and lumber out of the Eastern Oregon mountains until 1947, when Oregon Lumber Company replaced it with trucks. A short stretch of switching trackage in the Baker City mill remained until that closed in 1961.

A group of local enthusiasts started work on rebuilding part of the old Sumpter Valley Railroad for tourist trains in 1971, and they ran their first trains in 1976. The reconstruction to date has built about five and a half miles of track, most of that on the original SVRy grade. One of the principle locomotives used on the rebuild today is Heisler's construction number 1306, built new in 1915 for the W. H. Eccles Lumber Company as their #3. The Eccles company operated a mill served by the SVRy at Austin between 1911 and around 1925. The company then shifted its operations, including the #3, to Cascade, Idaho, but only carried on operations there for a year before selling out to Hallack and Howard Lumber Company. The H&H mill closed in 1930. The #3 went on to serve as a stationary boiler in the Boise-Cascade mill in Cascade, which ensured its survival until the SVRy restoration group could buy it.

MCEWAN, OREGON. Healthy lumber traffic levels in the middle 1910s prompted the Sumpter Valley Railway to buy five new Mikado type steam locomotives, three from Baldwin in 1915-1916, followed by two more from Alco in 1920. Alco built its construction numbers 61980 and 61981 as Sumpter Valley #101 and #102, but the railroad renumbered them #20 and #19 respectively shortly after their arrival. The pair worked twenty years on the SVRy until replaced by larger power in 1940.

The U.S. Army requisitioned the two locomotives along with other equipment for use on Alaska's White Pass & Yukon Railroad during World War II. The WP&Y assigned the pair numbers 81 and 80. The pair survived dieselization, and in 1977 the Sumpter Valley restoration group found them on the WP&Y dead line. They arrived back in Oregon in due time, and the SVRy group restored the #19 to operation in 1996. The #20 has been cosmetically restored and is stashed behind the SVRy shops in McEwan, Oregon, where it is a candidate for future restoration.

MINERAL, WASHINGTON. The Climax Locomotive Works built the third type of geared steam locomotives popular in the logging industry. Pennsylvania logger Charles Scott designed the locomotive, the later versions of which featured two cylinders inclined at a 22.5-degree angle driving a shaft underneath the boiler, which in turn powered another shaft running the length of the locomotive that powered all axles. The Climax Manufacturing Company of Corry, Pennsylvania, built approximately 1,035 locomotives between 1888 and 1928.

Climax built its construction number 1693 in March 1928 for the Hillcrest Lumber Company on Vancouver Island, British Columbia. The #10 worked there until 1968. The Mount Rainier Scenic Railroad bought the locomotive and brought it to Washington, where it remains in occasional operation.

GARIBALDI, OREGON. Another restored Heisler pulling tourists is construction number 1198, a 60-ton, 2-truck machine built in 1910 for the Curtiss Lumber Company of Mill City, Oregon, as their #2. The Heisler spent its entire career in Mill City, working for Hammond Lumber, Mill City Manufacturing, Vancouver Plywood & Veneer, and finally Willis Shingle Company. Jack Rogers moved the #2 to his Golden Age Museum in Ashford, Washington. Scott Wickert of Tillamook, Oregon, later bought the #2, and it was the first steam locomotive used on the Oregon Coast Scenic when it opened in 2003. It's been in operation on the railroad ever since.

CHEHALIS, WASHINGTON. Baldwin built its construction number 44106, a 90-ton logging Mikado, in 1916 new for the Clear Lake Lumber Company's Puget Sound & Cascade Railroad as its #200. Unfortunately, Clear Lake declared bankruptcy in 1926 before fully paying the #200 off, and Baldwin repossessed the locomotive.

Baldwin promptly resold the locomotive to the Cowlitz, Chehalis & Cascade Railway, who assigned it their #15. The CC&C dated from 1916, when it purchased the 10-mile long line of the Washington Electric Company running east from Chehalis, Washington. The company pushed the line farther east, reaching thirty-two miles to East Winston in 1928. Log traffic supported the CC&C until the railroad abandoned its line in 1955.

The CC&C donated the #15 to the City of Chehalis, who placed it on display. It would remain there until the Chehalis-Centralia Railroad Association obtained the locomotive in 1986 and hired the Mount Rainier Scenic Railroad shops restore it to operation. The organization ran its first trains in 1989. The now renamed Chehalis-Centralia Railroad & Museum continues to run excursions with the #15 on 10.2-miles of former Milwaukee Road/Chehalis Western/Curtis, Milburn & Eastern trackage running west from Chehalis that is owned by the Port of Chehalis.

WILLITS, CALIFORNIA. The Union Lumber Company incorporated the California Western Railroad & Navigation Company in 1905 to continue pushing a rail line east from Fort Bragg. The CWR&N reached Willits and a connection with the Northwestern Pacific Railroad in 1911. The railroad existed primarily to haul logs to Union's big Fort Bragg mill and then carloads of finished lumber to the NWP connection. The company dropped the "& Navigation" part of its name in 1947.

Passengers had always provided a small but steady stream of income, mostly provided by self- propelled railbuses. The railroad noticed an uptick in tourists riding its cars in the early 1960s, which caused the company to launch a full-fledged steam powered tourist train in 1965. Passenger revenues became increasingly important to the company, especially as freight traffic dwindled through the 1980s.

The California Western lost its freight business entirely in 1998 after the Federal Railroad Administration closed the NWP due to poor track conditions and other safety concerns. Georgia Pacific, corporate successor to Union Lumber, closed the Fort Bragg mill in 2003. The California Western has been beset by calamities ranging from tunnel collapses to storm damage to extensive manhunts for an armed suspect that have affected its operations, but it soldiers on, earning its entire livelihood hauling tourists through the redwoods out of Fort Bragg and Willits California Western #65, seen here in Willits, is a former Southern Pacific GP-9, built in 1954; it's been on the railroad since 1987.

4

MISCELLANEOUS

The book to this point has dealt mostly with physical manifestations of timber industry history. This last part deals with less tangible remnants, legacies, and tributes.

One of the cornerstones of timber industry propaganda in recent years is trees as a renewable resource. While this claim is entirely true, it's equally factual that the industry did not treat forests as renewable until the era after World War II. The geographic center of lumber production started in Maine in 1840 and migrated to the Upper Midwest over the next four decades, then shifted to the southern states, and finally to the Pacific Northwest. Concepts of reforestation and trees as a replenishable crop really didn't mature until after the industry reached the Pacific and suddenly had no more fresh uncut stands over the next hill. The wanderings of the industry resulted in the wholesale transplanting of names, social groups, and the like from region to region. It's hard to find a timber town in the northwest that doesn't have a street named Minnesota, Minneapolis, Michigan, Milwaukee, Saginaw, or some other reference to an Upper Midwest state or city. The former presence of the timber industry is marked in other ways—for instance, California alone has "Mill" attached to at least one hundred features, and "Sawmill" associated with over twenty more.

Another popular place to find timber industry tributes is in school or team mascots. Three universities—Northern Arizona in Flagstaff, Arizona; Humboldt State University in Arcata, California; and Stephen F. Austin State University in Nacogdoches, Texas, all use Lumberjack as their school mascot. Logger is a far more common mascot, used by the University of Puget Sound and a lengthy list of high schools, including Eureka and McCloud in California; Vernonia, Scio, and Mitchell in Oregon; Potlatch and St. Maries in Idaho; and many others. Professional sporting teams with timber industry inspired names includes baseball's Williamsport Crosscutters of Williamsport, Pennsylvania (Baseball, New York-Penn League) and Major League Soccer's Portland Timbers. The award for the most creative mascot almost certainly goes to the Grays Harbor College Chokers and their mascot Charlie the Choker, which references the logging occupation of choker setters.

The last part of this book will explore a few examples of these other traces of the timber industry, in art, names, and other forms.

TREES OF MYSTERY, CALIFORNIA. Paul Bunyan as the logging industry's mythological folk hero got its start in oral traditions in the logging camps of the Upper Midwest. The character may or may not have started with intertwined stories about two real French Canadians, a logger named Fabian "Joe" Fournier and a war hero named Bon Jean. The stories appear to have circulated through the camps for about thirty years before they first appeared in print around 1904.

The legend of Paul Bunyan really took off in 1914 when The Red River Lumber Company released the first of its advertisements drawn by William B. Laughead. The advertising campaign lasted until 1944 and consisted of many individual ads featuring stories about Bunyan and his exploits, plus periodic booklets. Laughead's work defined the proportions and general look of Bunyan and created many ancillary characters, including Babe the Blue Ox. Many of the stories Laughead used were original to him, while others were great embellishments of the older tales. Unfortunately, Laughead's work has become the basis for almost everything that is currently "known" about the character, and obfuscated much of the original tales.

Paul Bunyan has become a nearly universal presence across timber country. At least ten cities in Maine, Wisconsin, and Michigan claim to be his birthplace, and statues of Babe and Paul can be found almost everywhere trees grow. These statues of the pair, measuring 49 and 35 feet tall, greet drivers on Highway 101 to the Trees of Mystery attraction a few miles south of Crescent City, California.

BURNEY, CALIFORNIA. One of the often-overlooked aspects of the timber industry, especially in the pine forests of the West, was the role it played in developing California and parts of Arizona and Oregon into the agricultural powerhouses they became. Most of the markets for produce from those regions, especially citrus, lay in the east, and pine boxes were essential in getting the products delivered. Almost every sawmill in the pine country of Oregon, California, and Arizona operated a box plant adjacent to the mill, and quite often it would consume the bulk of the lumber the mill produced. In addition, lumber sales to stand-alone box factories accounted for a lot of pine lumber consumption.

The extent to which the California citrus growers became involved in the pine lumber industry underscored the importance of the relationship. The Sunkist Growers Cooperative organized Fruit Growers Supply Company in 1907 to stabilize their wood box supply, and through the years it operated sawmills in Hilt, Susanville, and Westwood, all in California. While it has largely been out of the manufacturing business for several decades, Fruit Growers remains one of the largest owners of private timberland in California, much of which is managed out of their office in Burney.

A partial list of other citrus companies once involved in the California lumber industry included the DiGiorgio Fruit Company (several sawmills in Quincy and elsewhere); California Fruit Exchange (Graeagle); and California Peach Growers Association (near Mather).

BEND, OREGON. The many statues of Paul Bunyan scattered around timber country aren't the only works of art remembering or paying tribute to the timber industry. Bend's transition from a timber town to a thriving tourist destination based in large part on outdoor recreation, breweries, and other pursuits has also spawned a thriving artist community, and some of their work pays homage to the history and heritage of the town. The Old Mill District is home to several sculptures, including the two seen here, one incorporating the flywheels from one of the Brooks-Scanlon sawmill's band saws and the other a pair of horses skidding a Ponderosa pine log composed of old metal pieces.

Two other sculptures in the immediate area of the Old Mill District feature old gears converted to raised planters and the crane boom and drag shovel once used to dredge the bed of the Deschutes River back when it served as a log pond. Scattered around the city are two sculptures of timber related professions, a logger bending down next to a stump and a planter with a pocket full of seedlings, and another horse figure wearing a collar set inside hoops inspired by the draft horses used in the logging industry in the region.

CROMBERG, CALIFORNIA. The Nibley-Stoddard Lumber Company, owned by two of the backers of the Oregon Lumber Company and Sumpter Valley Railway among a number of other businesses, built a small mill in Cromberg in 1923-1924. The mill and an associated logging railroad operated until a fire destroyed the mill on 29 May 1928. The company elected not to rebuild the mill after a fruitless attempt to purchase replacement sawmill equipment, and the logging railroad delivered loaded log cars to the Western Pacific for shipment to a mill in Loyalton, California. The effects of the Great Depression forced the operation to shut down in 1931.

Almost nothing remains of the Nibley-Stoddard company today. One of the few indicators a sawmill ever operated here is Old Mill Pond Road, which accesses several houses on and around the former mill site. It's typical of many such markers noting the former presence of a long-gone timber industry operation.

HINES, OREGON. Street names can be used to honour people as well as remember former homes in faraway places. An intersection of both can be found where Saginaw and Pettibone streets cross in Hines, Oregon. When the Edward Hines Lumber Company of Chicago won the massive Bear Valley Timber Sale in 1928 as related previously in this book, the company sent Frederick William Pettibone—better known as F. W.—west from its Mississippi operations to Oregon to oversee the herculean job of building and equipping the new venture. F. W. was the latest in a long line of Pettibones engaged in the lumber business, and the company chose well when it entrusted the job to him. Charles J. Pettibone, F. W.'s son and better known as C. J., followed his father into the business, and F. W. involved C. J. heavily in all of the key decisions. F. W. retired from the Hines company as the sawmill opened in 1930, but C. J. moved west to manage the operations his father built. C. J. presided over the trying years of the Great Depression and then guided the operations towards profitability before health concerns forced his retirement in 1942. The street named after the two men crosses another that harkens back to the upper Midwest roots of both the Hines company and many of the employees that followed the center of lumber production west.

ARCATA, CALIFORNIA. The State of California established Humboldt State Normal School as a teacher's college in 1913. The institution changed names several times, first to the Humboldt State Teacher's College and Junior College (1921), the Humboldt State College (1935), California State University, Humboldt (1972), and finally to Humboldt State University (1974). Humboldt named its sporting teams the Thunderbolts until 1936, when the school changed to the Lumberjacks. The school adopted Lucky Logger as its mascot in 1959.

In 1968, a group of students reformed the school's long-gone marching band as the Humboldt State Marching Lumberjacks, usually shortened to MLJs. The band is conducted by an axe major using an axe instead of a drum major, and the band uniform consists of work boots, green pants, yellow band shirts, suspenders, and a yellow hardhat. The often irreverent MLJs have always been student run and directed, and in addition to school sporting events, the band marches in parades up and down the length of California. The band celebrated its fiftieth anniversary in November 2018 when 300 current and former band members gathered and performed at what is likely to be Humboldt's final ever home football game. The band is seen here performing in downtown Arcata on Halloween 2018.

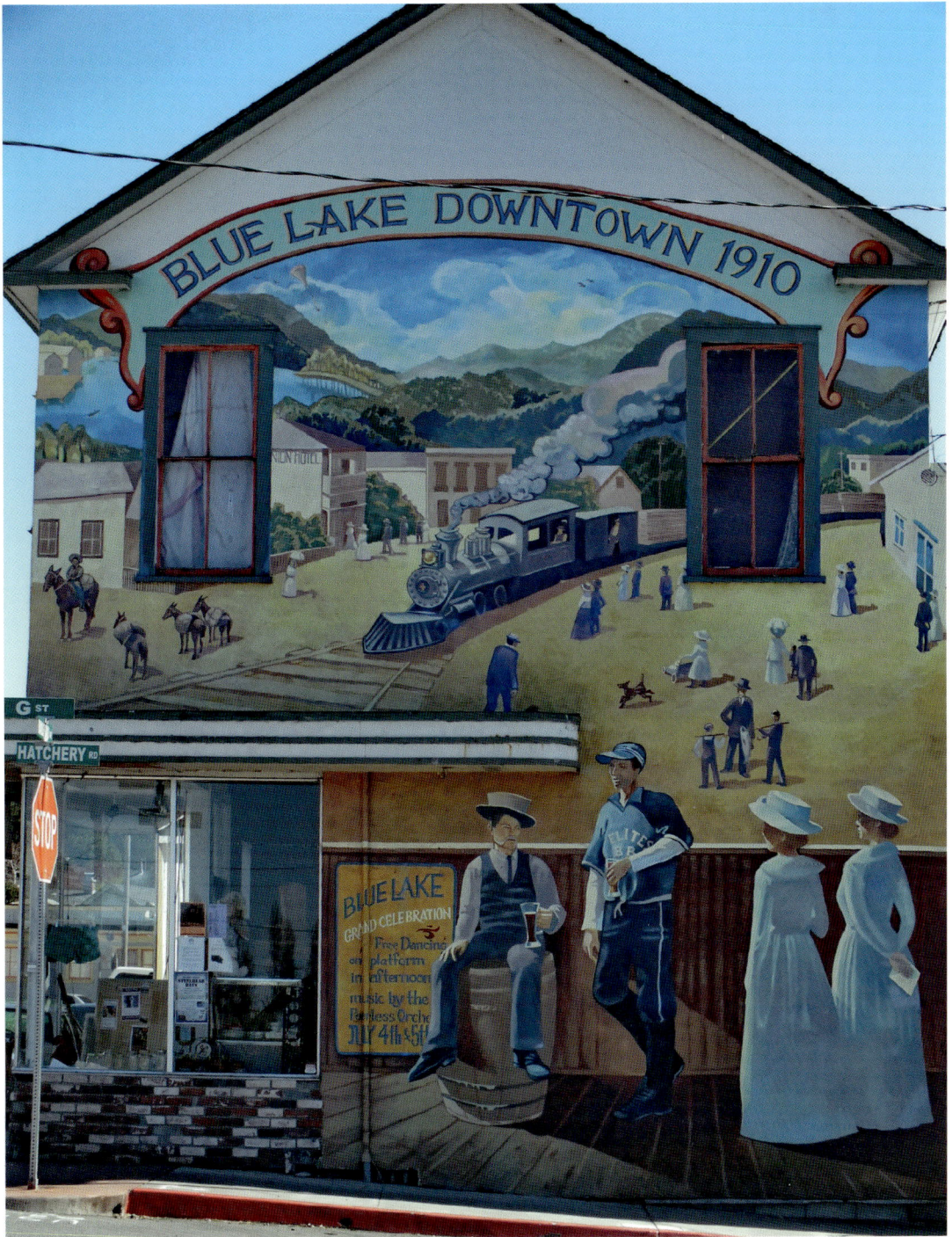

BLUE LAKE, CALIFORNIA. Scenes of logging and lumbering heritage are quite often depicted in murals scattered throughout timber country. Centerpiece of this mural on the side of a building in Blue Lake depicting an idyllic summer day in 1910 is an Arcata & Mad River Railroad narrow gauge train bringing another load of lumber from the mill in Korbel and headed for the wharf in Arcata. The A&MR grade is just out of view to the right, and the mural faces the old A&MR depot that now houses the local museum.

FLAGSTAFF, ARIZONA. A retaining wall next to the Atchison, Topeka & Santa Fe passenger depot in Flagstaff hosts an extensive mural featuring highlights of the region's history and economy. The lumbering heritage is depicted by this set of horses and high wheels skidding logs out of the forests. Similar murals can be found elsewhere, such as in Susanville, California.

BURNEY, CALIFORNIA. In the early 1990s the Burney Chamber of Commerce collaborated with local businesses and the Burney Jr. Sr. High School's art department to create a number of murals around town. The art teacher, well aware of this author's interest in railroads, invited me to help paint this mural, featuring a McCloud River Railroad log train and applied to the side of the bowling alley. An addition to the building has cut part of the mural off, but save for some graffiti, the rest of the painting is still in decent shape roughly twenty years after it was painted.

SWEET HOME, OREGON. Nestled next to the South Santiam River on the western flank of the Cascade Mountains, Sweet Home is another town owing its existence to the timber industry. Unlike many other towns, Sweet Home still has a few active forest products plants, but the remaining industry is still but a shadow of what it used to be, and the shuttered and abandoned mill sites far outnumber the still active ones. The town is still justifiably proud of its history, as evidenced by this mural adorning most of the side of a building just off Highway 20 near the middle of downtown. What sets this mural apart from many others is its attention to the human and animal details of the industry in the days before chainsaws, hard hats, internal combustion engines, and the other tools of the modern trade.

McCLOUD, CALIFORNIA. Many current and former timber communities have been around long enough to start marking major milestones in recent decades. McCloud celebrated its 100th anniversary in 1997 with a year's worth of events. The town created this logo for the occasion, featuring a logger, waterfall, Mount Shasta, and the McCloud River Railroad #25 powering a log train. The Heritage Junction Museum of McCloud has incorporated the logo into the sign for their museum building.

CHILOQUIN, OREGON. The collection of the Collier Logging Museum includes a lot more than just equipment. Other tributes to the industry on the grounds include a pair of cork boots, a hardhat, and a logger's lunch box, all beautifully carved in wood. The museum elsewhere has one of the largest collections of preserved chainsaws anywhere. It's a fascinating place.

KLAMATH FALLS, OREGON. A former Oregon, California & Eastern Railroad caboose on display at the OC&E Woods Line trailhead in Klamath Falls.

SELECTED BIBLIOGRAPHY
AND REFERENCES

Adams, Kramer. *Logging Railroads of the West.* Bonanza Books, New York, NY, 1961.

Bowden, Jack. *Railroad Logging in the Klamath Country.* Oso Publishing Company, Hamilton, MT, 2003.

Bretz, Berden O. *The Lonesome Whistle.* Crescent City Printing Company, Crescent City, CA, 1993.

Carranco, Lynwood and John T. Labbe. *Logging the Redwoods.* The Caxton Printers, Ltd., Caldwell, ID, 1989.

Carranco, Lynwood and Henry L. Sorenson. *Steam in the Redwoods.* The Caxton Printers, Ltd., Caldwell, ID, 1988.

Johnson, Merv. *In Search of Steam Donkeys.* Timber Times, Hillsboror, OR, 1996.

Moore, Jeff. *California's Lumber Shortline Railroads* and *Eastern Oregon Shortline Railroads.* Both Fonthill Media, Charleston, South Carolina, 2016.

The McCloud River Railroads. Signature Press, Winton, CA, 2016.

Myrick, David F. *Railroads of Nevada and Eastern California, Volume III: More on the Northern Roads.* University of Nevada Press, Reno and Las Vegas, NV, 2007.

Tedder, Russell. *Forest Rails, Georgia-Pacific's Railroads.* White River Productiosn, Bucklin, MO, 2016.

Young, James A. and Jerry D. Budy. *Endless Tracks in the Woods.* Originally published by Crestline Publishing in 1989, then republished by Motorbooks International, Osceola, WI, 1993.

Various periodicals including *Railfan & Railroad, Timber Times, Tall Timber Short Lines, Timberbeast,* and others.